Graphis Packaging 8

Graphis Packaging 8

Graphis Packaging 8

An International Compilation of Package Design

Ein Internationaler Überblick über die Packungsgestaltung

Un Repetoire International des Formes d'Emballage

Publisher and Creative Director: B. Martin Pedersen

Editors: Heinke Jenssen
Art Director: Lauren Slutsky

Assistant Editor: Michael Porciello
Design & Production Assistant: Joseph Liotta

Published by Graphis Inc.

(opposite and following page) Tanqueray #10 Gin bottle by Deutsch, Inc.

Contents Inhalt Sommaire

Commentary by Steve Sandstrom 6

Beverages, Getränke, Boissons 16

Communication, Kommunikation, Communication 82

Cosmetics, Kosmetik, Cosmétiques 86

Fashion, Mode, Mode 112

Food, Nahrungsmittel, Alimentation 118

Games, Spiele, Jeux 156

Household, Haushalt, Articles Ménagers 158

Industrial, Industrie, Industrie 176

Multimedia, Multimedia, Multimedia 182

Museums, Museen, Musées 188

Music CDs, Musik-CDs, CD 190

Paper, Papier, Papiers 192

Pharmaceuticals, Pharmaka, Produits pharmaceutiques 198

Promotion, Promotionen, Promotions 206

Shopping Bags, Tragtaschen, Sacs à Main 220

Sports, Sport, Sport 230

Tobacco, Tabakwaren, Tabacs 236

Toys, Spielzeug, Jouets 244

Indices, Verzeichnisse, Index 249

Remarks: We extend our heartfelt thanks to contributors throughout the world who have made it possible to publish a wide and international spectrum of the best work in this field. Entry instructions for all Graphis Books may be requested from: **Graphis Inc.**, 307 Fifth Avenue, Tenth Floor, New York, NY 10016 or visit our Web site, www.graphis.com

Anmerkungen: Unser Dank gilt den Einsendern aus aller Welt, die es uns durch ihre Beiträge ermöglicht haben, ein breites, internationales Specktrum der besten Arbeiten zu veröffentlichen. Teilnahmebedingungen für die Graphis-Bücher sind erhältlich bei: **Graphis Inc.**, 307 Fifth Avenue, Tenth Floor, New York, NY 10016. Besuchen Sie uns im World Wide Web, www.graphis.com

Remerciements: Toute notre reconnaissance va aux designers du monde entier dont les envois nous ont permis de constituer un vaste panorama international des meilleures création. Les modalités d'inscription peuvent être obtenues auprès de: **Graphis Inc.**, 307 Fifth Avenue, Tenth Floor, New York, NY 10016. Rendez-nous visite sur notre site web: www.graphis.com

SS-01

SS-15

SS-15

SS-01

Steve Sandstrom on Packaging

My deodorant has racing stripes on it. It looks more like a product that I might have purchased at an auto parts or sporting goods store instead of at the supermarket. I hadn't given much thought to these racing stripes before. I'm not exactly sure why this motif emblazons the face of this particular product, but it must have something to do with appealing to my masculine subconscious. Maybe the stripes are meant to represent or inspire an active lifestyle. An active lifestyle could create more perspiration, thus creating the need for more deodorant. Maybe the stripes are just an icon of *machismo*—a quick glance will let you know this stuff is for real men. *Fast men.* There is even a brand of deodorant called Speed Stick. Personally, these racing stripes and speed references don't make me or my armpits feel any faster.

Many personal products for men are packaged with a similar approach—shaving products, in particular. From razor blades to shaving foam, racing stripes and other designs of speed seem prevalent. I wonder if anyone in the deodorant business has ever questioned this design solution or considered any other options.

It is almost impossible to find an aesthetically pleasing package in the personal products for men sections of supermarkets or drugstores. Batteries could be sold alongside these products because battery packaging is almost identical in its approach. It may not be fair to criticize these products alone for their packaging, because I think the majority of all products in the supermarket are packaged with tired concepts, busy layouts, odd graphic embellishments, poorly executed identities and confused combinations of typefaces. I know of countless graphic designers who could create better-looking packaging solutions for deodorant than what exists in the stores today. In fact, I believe it could be done in less than a day. This may not be enough time for a talented designer to do his or her very best concept, but it is enough time to do something better. If the goal is to appeal to men by looking sporty, imagine what a deodorant from a design-savvy company like Nike might look like.

I asked a friend who lives in Tokyo if deodorant was packaged differently in Japan than in the U.S.. The Japanese have done so many beautiful things with packaging, I thought it might be inspiring to see their approach. After some research, he sent me digital images of several brands sold there. Many of these brands were the same as in America, with essentially the same design. Others were the same brands, but they were completely redesigned from the container shapes to logos and colors. And although these Japanese products looked better than most sold in this country, there was still much room for improvement.

There's no reason why deodorant packaging for men shouldn't be better. I have seen and purchased shampoos, lotions, soaps and other personal products from specialty shops that were unusual in their approach to package design and great to look at. Most of these products were from England or France—only a few were from America. I wanted and admired them for their style and color, their interesting use of type and label design, their creative use of materials and closures, and their overall sophistication. Some looked "old world" and others looked very new. Each was able to charm me into thinking that they were of high quality and that if I owned them I might be someone special. And perhaps the most intelligent and compelling effect created by each of these package designs is the feeling that these products were made by people and not by corporations. It seems only appropriate that personal products should feel personal.

I am sure that there has been extensive consumer and marketing research to prove racing stripes would be an exceptional graphic solution for men's deodorant. There are likely to be marketing executives who have moved up the corporate ladder for having successfully launched a new, sportier look for their line of regular and unscented stick (or *schtick* as it may be). But all we usually get out of big packaged goods companies' marketing research and focus groups is uninspired packaging. Hey, I bought it. But, I didn't really feel like I had much choice. And like I said, I haven't really given much thought to deodorant packaging before. Which is probably how much anyone cares about deodorant packaging to begin with. All anyone wants is dry armpits. Who cares what the package looks like as long as it works. Or in most cases, kind of works.

There is nothing repulsive about the deodorant packaging on the retail shelves today. Nothing that looks so terrible or threatening that it keeps one from purchasing any of it. There is nothing really good about it either. It is all about the same. Which made me think nothing of it for years. I'm certainly not very loyal to any brand and I hold the assumption that they are all about the same product with only a scent making the difference. And since I usually buy the unscented solid type, I'll buy any brand available in that form. I believe this kind of packaging and perception of product parity can be an opportunity for a brand to use design as a tremendous marketing weapon to help separate one product from the rest. Even if all it does is make it look better. In this case, that would be a significant difference.

When a company adopts the strategy to apply thoughtful design to their packaging it can improve a brand's image and market position by projecting its uniqueness and quality and by positively affecting how people feel and respond toward it. And sometimes it can create a shift in the entire category. Think about California wines and the amount of effort spent to create more attractive and unusual label designs over the last two decades. Many of the early adapters were denounced as "designer" wines. They may not have been the best wines, but they sold well. And that's usually enough to start a movement. Now there is hardly a winery of any quality in California which hasn't put forth an attempt to use better packaging. We've seen a few good design solutions for olive oil, tea and micro-brewed beers put pressure for better packaging on all designers of such products. Shouldn't deodorant aspire to do better too? Isn't it time for deodorant to raise up its ugly self and move beyond the cotton swabs and dental floss, out of the shame of dark cabinets and drawers and on to the open bathroom counter of respect next to that fine cologne? I say it's time for designer deodorant. And then it's onward to the exclusive shelves of airport duty free stores. Let the shaving foam follow.

Brand Identity for Tazo Teas by Sandstrom Design

File No. T47-002a:

TAZO COASTERS

Steve Sandstrom: Reflexionen über Verpackungen

Auf meinem Deodorant sind Rennstreifen. Er sieht eher wie ein Produkt aus einem Autozubehör- oder Sportgeschäft aus als wie ein Artikel aus dem Supermarkt. Ich habe mir eigentlich nie zuvor Gedanken über diese Rennstreifen gemacht. Mir ist nicht ganz klar, warum dieses Motiv dieses bestimmte Produkt ziert, aber es muss damit zu tun haben, dass es das männliche Unterbewusstsein ansprechen soll. Vielleicht sollen die Streifen einen aktiven Lebensstil darstellen oder zu einem solchen anregen. Ein aktiver Lebensstil würde mehr Transpiration bedeuten und somit den Verbrauch von Deodorant fördern. Vielleicht sind die Streifen auch nur ein Zeichen von *Machismo* – ein schneller Blick auf das Produkt, und man weiss sofort, dass es für richtige Männer ist. *Für schnelle Männer.* Es gibt sogar eine Deodorant-Marke, die «Speed Stick» heisst. Mir persönlich geben diese Rennstreifen und Andeutungen von Geschwindigkeit nicht das Gefühl, schneller zu sein.

Viele Artikel für die persönliche Pflege des Mannes sind ähnlich verpackt – besonders Rasierartikel. Rennstreifen dominieren das Bild. Ich frage mich, ob irgendjemand im Deodorant-Geschäft dieses Design je in Frage gestellt oder andere Gestaltungsmöglichkeiten in Betracht gezogen hat. Es ist fast ausgeschlossen, unter den Pflegeprodukten für Männer in Gemischtwarenläden oder Drogerien ansprechende Verpackungen zu entdecken. In der Nachbarschaft dieser Artikel sind Batterien durchaus nicht fehl am Platz, denn ihre Verpackung sieht fast genauso aus. Es ist vielleicht nicht fair, diese Artikel wegen ihrer Verpackung herauszupicken, denn ich bin überzeugt, dass die Mehrheit aller im Supermarkt angebotenen Produkte nach demselben überstrapazierten Konzept verpackt sind und überladene Layouts, seltsame Verzierungen, schlecht umgesetzte Markenzeichen und hilflose Kombinationen von Schriften aufweisen. Ich kenne zahlreiche Graphik Designer, die bessere Gestaltungslösungen für Deodorants finden könnten. Sie würden meiner Meinung nach nicht einmal einen Tag dazu brauchen. Das ist vielleicht für einen begabten Designer nicht genug Zeit, um ein optimales Konzept auszuarbeiten, aber es wäre genug Zeit, etwas Besseres zu machen. Wenn das Ziel ist, durch einen sportlichen Look Männer anzusprechen, stelle man sich einmal vor, wie ein Deodorant von einer Design-bewussten Firma wie Nike aussehen würde.

Ich fragte einen Freund in Tokio, ob Deodorants in Japan anders aussehen als in den USA. Die Japaner haben so viele schöne Dinge mit Verpackungen gemacht, und ich dachte, es könnte sehr anregend sein, ihre Lösungen zu sehen. Nach einigen Nachforschungen seitens meines Freundes erhielt ich digitale Bilder von verschiedenen in Japan angebotenen Deodorant-Marken. Viele davon waren dieselben, die man auch in den USA bekommt, und ihre Verpackung unterschied sich kaum. In anderen Fällen waren es zwar dieselben Marken, aber sie waren völlig anders gestaltet als die bei uns angebotenen Artikel. Diese Artikel sahen zwar besser aus als unsere, aber auch für sie hätte sich noch viel tun lassen.

Es gibt keinen Grund, warum Deodorants für Männer nicht besser aussehen könnten. Ich habe in Spezialgeschäften Shampoos, Lotionen, Seifen und andere Pflegeartikel mit ungewöhnlichen Verpackungen entdeckt. Die meisten kamen aus England oder Frankreich und nur ganz wenige aus Amerika. Mir gefielen Stil und Farbe, Typographie und Design der Etiketten, Material und Verschlüsse, Kreativität und Raffinesse der gesamten Verpackung. Einige sahen nach der «alten Welt» aus, andere wirkten sehr modern. Jedes einzelne Produkt sprach mich an und liess mich glauben, dass es von hervorragender Qualität sei und dass ich, wenn ich es besitzen würde, jemand Besonderes sei. Das Wichtigste und Intelligenteste daran war vielleicht der von all diesen Artikeln vermittelte Eindruck, dass sie von Menschen hergestellt wurden und nicht von Grossunternehmen. Es ist naheliegend, dass Artikel für die persönliche Pflege auch persönlich aussehen sollten.

Die Deodorants, die man heute in den Ladenregalen findet, sehen nicht abstossend aus, auf jeden Fall nicht so fürchterlich und beängstigend, dass man sie nicht kaufen würde. Aber es gibt auch nichts Schönes an ihnen zu entdecken. Sie sehen alle ziemlich gleich aus. Darum habe ich auch all die Jahre keinen Gedanken an ihr Aussehen verschwendet. Ich habe keine besondere Vorliebe für eine bestimmte Marke, weil ich überzeugt bin, dass sich die Produkte in ihrer Wirkung kaum unterscheiden, sondern nur durch die Duftnote. Da ich eine unparfümierte Sorte benutze, kaufe ich, was es in dieser Art gerade gibt. Meiner Meinung nach liegt hier angesichts der vorhandenen Verpackungen und der festzustellenden Übereinstimmung der Produkt-eigenschaften eine grosse Chance für eine Firma, Design sehr erfolgreich als Marketinginstrument einzusetzen, um ihr Produkt von allen anderen vorteilhaft abzusetzen. Selbst wenn man nur das Äussere des Produktes verschönern würde – in diesem Fall wäre das ein entscheidender Unterschied.

Wenn eine Firma sich zu einem gut durchdachten Packungsdesign ihrer Produkte entschliesst, kann sie Image und Marktposition des Produkts durch Betonung seiner Einmaligkeit und Qualität verbessern. Manchmal kann sie dadurch eine Veränderung in einer ganzen Produktkategorie bewirken. Man denke nur an die Weine aus Kalifornien und an die in den letzten zwanzig Jahren unternommenen Bemühungen um schönere und speziellere Etiketten. Viele der ersten Anstrengungen in dieser Richtung führten zu der Bezeichnung "Designer-Weine". Es waren vielleicht nicht die besten Weine, aber sie verkauften sich gut. Meistens ist das genug, um etwas zu bewegen. Heute gibt es kaum ein Weingut in Kalifornien, das sich nicht um eine bessere Verpackung seiner Erzeugnisse bemüht hat. Dasselbe ist bei Olivenöl, Tee und Bier von Keinstbrauereien passiert – gute Verpackungen einiger Erzeugnisse haben dazu geführt, dass die Verpackungen aller Produkte dieser Kategorien besser wurden. Sollten die Deodoranthersteller nicht daraus lernen? Ist die Zeit nicht reif für Deodorants, ihr hässliches Äusseres, ihr Schattendasein in dunklen Schubladen und Badezimmerschränken aufzugeben und sich einen Platz auf den Tablaren zu erobern? Ich sage, die Zeit ist reif für Designer-Deodorants. Alles Nächstes lasse man bitte den Rasierschaum folgen.

Steve Sandstrom: Réflexion sur les emballages

Mon déodorant, qui présente les mêmes traînées que celles laissées par les pneus d'un bolide, semble plutôt provenir d'un magasin d'accessoires de voiture ou d'une boutique d'articles de sport que du supermarché. Je n'avais encore jamais vraiment prêté attention à ces traînées. J'ignore pourquoi ce motif a été choisi pour le conditionnement de ce produit, mais il n'est pas impossible qu'il soit censé titiller le subconscient masculin. A moins que la fonction de ces traînées ne soit de symboliser ou de préconiser un style de vie dynamique, car l'activité stimule la transpiration et, partant, la consommation de déodorant. Ou peut-être l'idée était-elle de puiser dans le registre machiste: un seul regard suffit en effet pour se rendre compte que le produit s'adresse à des hommes virils, vifs et rapides. Il existe d'ailleurs un déodorant de la marque Speed Stick. Personnellement, je dois avouer que ces traînées évanescentes et ces références à la vitesse ne me donnent nullement l'impression d'être plus rapide. Les packagings de produits pour homme conçus dans le même esprit ne manquent pas, notamment pour le rasage. Sur les sachets de lames de rasoir, les tubes de gel ou les bombes de mousse à raser, les traînées évanescentes et les symboles de vitesse sont des motifs récurrents. Je serais curieux de savoir si, dans le segment du déodorant, quelqu'un a jamais remis ces solutions graphiques en question ou pris d'autres options en considération.

Dans les rayons de produits pour homme d'un supermarché ou à la droguerie du coin, inutile de chercher un emballage attractif. Les piles pourraient aussi bien être rangées au même endroit sans déparer le moins du monde les linéaires. Peut-être n'est-il pas de bonne guerre de critiquer l'aspect extérieur de ces articles en particulier car, à mon avis, les grandes surfaces n'ont quasiment rien d'autre à proposer en la matière que des concepts éculés, des présentations surchargées, des fioritures graphiques hideuses, des identités visuelles bâclées et des combinaisons typographiques maladroites. Parmi mes connaissances, plus d'un designer graphique serait capable de créer un habillage plus attrayant des déodorants que tout ce qu'on a pu voir à ce jour. Et je suis persuadé qu'il leur faudrait moins d'une journée pour arriver à ce résultat. Dans ce délai, les plus doués ne pourraient peut-être pas aller jusqu'au bout de leurs possibilités, mais encore une fois, le progrès serait flagrant. Si le but est d'interpeller les hommes en jouant la carte du look sportif, imaginez à quoi ressemblerait le déodorant d'une marque aussi portée sur le design que Nike.

J'ai demandé à un ami qui habite Tokyo si les déodorants étaient mieux là-bas qu'aux Etats-Unis. Les Japonais étant passés maîtres dans l'art de l'emballage, je me disais qu'il serait intéressant d'examiner leur approche. Après avoir mené sa petite enquête, il m'a envoyé des images numériques de produits de différentes marques. Certains ne présentaient aucune différence avec nos articles tandis que d'autres, de même marque, avaient été entièrement revisités. S'ils étaient nettement plus attractifs que ceux que l'on trouve ici, bien des progrès restent encore à faire. Rien n'empêche d'améliorer les déodorants pour hommes. J'ai déjà vu et même acheté, dans des boutiques spécialisées, des produits de toilette originaux et agréables à regarder, tels que shampooings, lotions ou encore savons. La plupart d'entre eux venaient d'Angleterre ou de France, rarement des Etats-Unis. J'ai aimé leur style, leur couleur, leur logo, leur texture, leurs systèmes de fermeture et toute la recherche que l'on sentait derrière. Certains avaient un air rétro, d'autres plutôt avant-gardiste. S'ils m'ont conquis, c'est parce que j'avais l'impression qu'ils étaient d'excellente qualité et qu'en me les appropriant, je ne serais plus Monsieur lambda. Mais le principal attrait est qu'ils semblaient avoir été créés par des individus et non par de grandes entreprises. N'est-il pas normal, en effet, que les produits de toilette donnent l'impression d'être personnalisés?

Je suis sûr que de vastes études de marché ont conclu que les traînées évanescentes étaient LA solution graphique pour les déodorants pour hommes. Des responsables du marketing ont probablement gagné de l'avancement pour avoir dopé la vente de leur gamme de sticks inodores en leur donnant un look plus sportif. Mais force est de constater que les études de marketing réalisées par les grandes entreprises sur les emballages ne débouchent généralement que sur des résultats insipides.

Les déodorants que l'on trouve actuellement dans le commerce n'ont rien de répugnant. Ils ne sont pas horribles ou menaçants au point d'exercer un effet dissuasif sur le consommateur. Ils n'ont rien de vraiment attirant non plus. Ils se caractérisent par leur uniformité. C'est d'ailleurs ce qui explique que j'y sois resté indifférent durant des années. Je ne suis fidèle à aucune marque en particulier. A mes yeux, elles se valent toutes et elles se distinguent uniquement par leur odeur. Comme je préfère les sticks inodores, c'est selon ce critère que je les achète. A mon avis, la marque qui serait capable de retravailler son emballage de manière à différencier son déodorant de ceux de ses concurrents se doterait d'un puissant outil de marketing. Un produit agréable à regarder, voilà qui ferait la différence.

Une compagnie qui soigne ses emballages peut améliorer son image de marque et sa position sur le marché en faisant ressortir la qualité et le caractère unique de son produit. L'emballage conditionne la perception du consommateur à l'égard du produit. Et cette innovation peut avoir des répercussions sur tous les produits de la même catégorie. Les efforts déployés ces vingt dernières années pour réinventer les étiquettes des vins californiens en est un bon exemple. Cela leur avait d'ailleurs valu l'appellation de «vins de designers». Ce n'étaient peut-être pas les meilleurs, mais ils se sont bien vendus. Et il n'en faut souvent pas plus pour amorcer une tendance. Depuis, quasiment tous les viticulteurs californiens surveillent leur identité visuelle. Les fabricants d'olive, d'huile, de thé et de bière leur ont emboîté le pas. Cela ne devrait-il pas inciter les fabricants de déodorant à en faire de même? A faire en sorte que leurs produits quittent l'obscurité des fonds de tiroirs où on les relègue pour rejoindre les rayonnages des salles de bain où trônent fièrement les beaux flacons de parfum? J'affirme que l'heure des "déodorants design" a sonné. Rien ne s'opposera alors à ce qu'ils partent à la conquête des boutiques d'articles hors taxe des aéroports. Et il sera temps, dès lors, de s'occuper de la mousse à raser.

Graphis Packaging 8

Agency: Work, Inc. Creative & Art Director: Cabell Harris Designer: Haley Johnson Photographer: Karl Steinbrenner Client: Work Beer

Agency: Turner Duckworth Creative Director: David Turner, Bruce Duckworth Designer: David Turner Client: McKenzie River Corp.

Agency: Deutsch Design Works Creative Director: Barry Deutsch Designer: Lori Wynn Illustrator: Larry Duke Client: Great Northern Brewing Co

Agency: Parachute Design Creative Director: Jac Coverdale Art Director & Designer: Bob Upton Illustrator: Peter Krause Copywriter: Kelly Trewartha Client: Clarity Coverdale Fury

(top) Agency: Deutsch Design Works Creative Director: Barry Deutsch Designer: Lori Wynn, Jacques Rossouw Illustrator: Jeff Norwell Client: Anheuser-Busch (bottom) Agency: Deutsch Design Works Creative Director: Barry Deutsch Designer: Lori Wynn Illustrator: Taylor Oughton, Greg Beecham Client: Anheuser-Busch

Agency: Mark Oliver, Inc. **Creative & Art Director, Copywriter:** Mark Oliver **Designer & Illustrator:** Mark Oliver, Patty Driskel **Client:** Firestone/Walker

Agency: Sibley Peteet Design Creative Director: Don Sibley Art Director & Designer: Tom Hough Illustrator: Calef Brown Client: Gambrinus Company

Design, Inc. Creative Director, Designer & Illustrator: Mary Anne Mastandrea Photographer: Erik Butler Copywriter: Lane Foard Client

ncy: Cornerstone Creative Director: Keith Steimel Art Director: Sally Clarke Designers: Joe Dimeo, Sally Clarke, Keith Steimel Illustrators: Clint Hansen, David O'Neal Client: Twisted wing Company

Agency: Shin Matsunaga Design, Inc. Creative & Art Director, Designer: Shin Matsunaga Client: Takara Shuzo Co., Ltd.

Agency: Caldewey Design Creative & Art Director, Designer: Jeffrey Caldewey Client: Revolution Hardrinks

Agency: Bill Carson Design Creative & Art Director, Designer, Illustrator: Bill Carson Client: Mountain Sun Organic Juices

Agency: Logos Identity by Design Ltd. Creative & Art Director: Brian Smith Designer: Gabriella Sousa Photographer: Lowthar Ulrich Client: The Great Atlantic & Pacific Co. of Canada, Ltd

Agency: Blackburn's Ltd. Creative & Art Director: John Blackburn Designer: Sarah Roberts Client: Orchid Drinks

Agency: Lloyd Ferguson Hamlins **Creative Director:** Mark Lloyd **Art Director:** Tony Enoch **Designer:** Tony Enoch, Steve Irvine **Illustrator:** Bob Stradling **Copywriter:** Richard Patteson **Client:**
Greater Europe

Agency: Mario Milostic Design Creative & Art Director, Designer, Copywriter: Mario Milostic Client: Vinograd Croatia

Agency: Van Noy Group **Creative Director:** Jim Van Noy **Art Director:** Bill Murawski **Designer:** Amanda Park **Photographer:** Lefteris Padavos **Illustrator:** Cat Landry **Client:** Gaetano Specialties,

Agency: Klim Design Creative & Art Director, Designer: Matt Klim Photographer: Greg Klim Client: Jose Cuervo

Agency: Van Noy Group **Creative Director:** Jim Van Noy **Art Director:** Bill Murawski **Designer:** Bill Murawski, Amanda Park **Photographer:** Lefteris Padavos **Illustrator:** Bob Schuchman **Cl**
Gaetano Specialties, Ltd.

Agency: Klim Design Creative & Art Director, Designer: Matt Klim Photographer: Greg Klim Client: Jose Cuervo

Agency: Deutsch Inc. Creative Director: Craig Markus Designer: Craig Markus, Genevieve Gorder Client: Tanqueray

(above) Agency: Scandinavian Design Group Designer: Morten Fornebo, Kristine Lillerik Illustrator: Arild Sater Client: Arcus Produkter AS (opposite) Agency: Spar Advertising Creative Director: Lane Casteix Art Director: Catherine Corley-Macacy Designer: Catherine Corley-Macacy Client: Sazerac Company

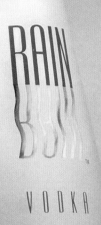

Agency: Ceradini Design, Inc. Creative Director: David Ceradini, Lori Raymer Designer: Keri Piatek, Lori Raymer Illustrator: Kathy Petrauskas Client: United Distillers & Vintners North America

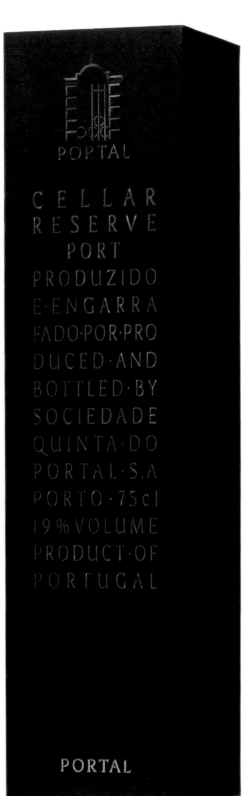

Agency: Harcus Design Art Director: Annette Harcus Designer: Hans Kohla Photographer: Michael Donavan Copywriter: Nicole Potter Client: Yalumba Winery

Agency: Harcus Design Art Director: Annette Harcus Designer: Melonie Ryan Photographer: Keith Arnold Illustrator: Melonie Ryan Client: Tallara Wines

On label: 1996 Sonata A BLEND OF 80% CABERNET SAUVIGNON AND 20% MERLOT Livermore Valley WENTE VINEYARDS

Agency: Michael Osborne Design Creative Director: Michael Osborne Designer: Paul Kaginada Illustrator: Caio Fonseca Client: Wente Vineyards

Agency: Britton Design Creative & Art Director, Designer, Illustrator: Patti Britton Client: Viansa Winery

Agency: Britton Design Creative & Art Director, Designer, Illustrator: Patti Britton Client: Buena Vista Winery

Agency: Acme Graphic Design Art Director & Designer: Greg Holly Client: Duck Pond Cellars

Agency: Blackburn's Ltd. Creative & Art Director: John Blackburn Designer: Roberta Oates Illustrator: Martin Leman Client: Sociedade Quinta Do Portal SA

Agency: Louise Fili, Ltd. Creative & Art Director: Louise Fili Designer: Louise Fili, Mary Jane Callister Client: Matt Brothers

Agency: Deutsch Design Works **Creative Director:** Barry Deutsch **Designer:** Barry Deutsch, Dawn Janney **Illustrator:** Jody Howgil **Client:** Robert Mondavi Winery

Agency: Packaging Create, Inc. Art Director: Aiko Okumura Designer: Koichi Hishida Client: Gekeikan, Inc.

Agency: Blackburn's Ltd. Creative & Art Director: John Blackburn Designer: Belinda Duggan Illustrator: James Marsh Client: Sociedade Dos Vinhos Borges

Agency: Pentagram Design Creative & Art Director: Kit Hinrichs Designer: Jackie Foshaug Illustrator: Daniel Pelaun Client: Swiss Dairy

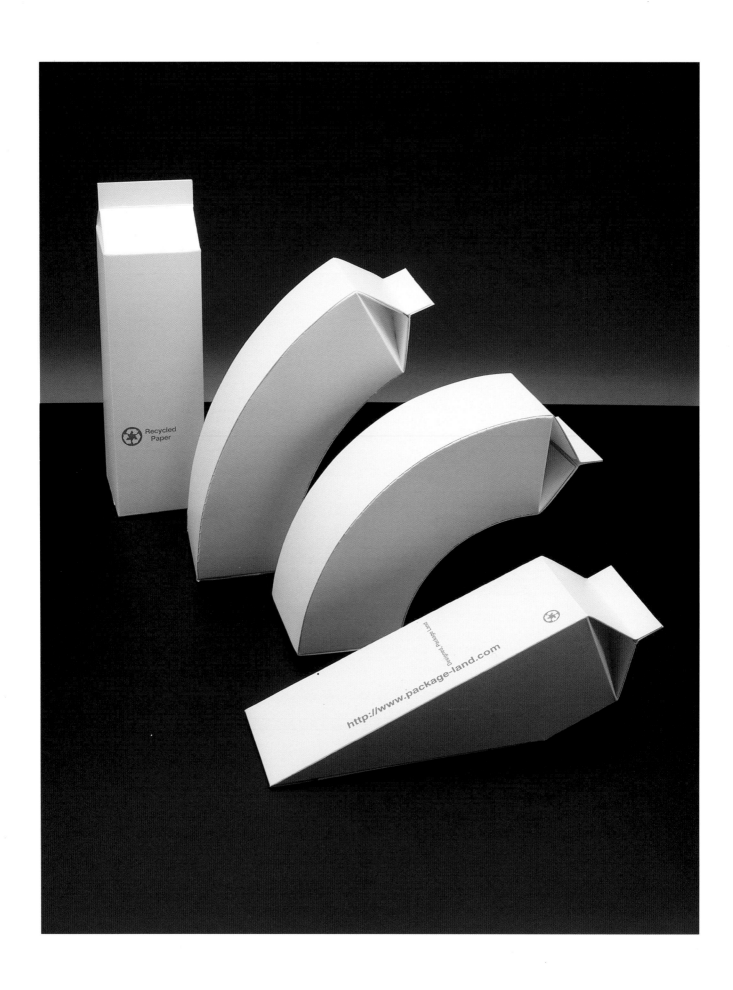

Recycled Paper

http://www.package-land.com

Designed Package Land

Agency: Guld & Grüna Skogar Design Creative Director: Jörgen Olofsson Designer: Mats Svensson, Håkan Olofsson Client: Skånmeerier

Agency: Blackburn's Ltd. Creative & Art Director: John Blackburn Designer: Roberta Oates Client: Orchid Drinks

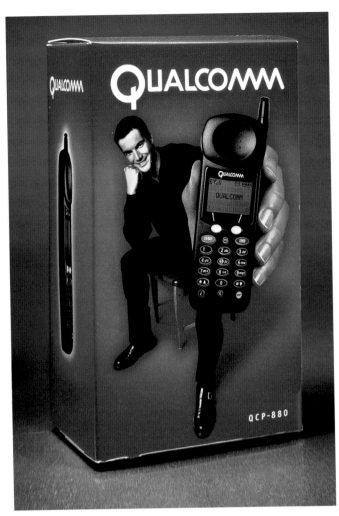

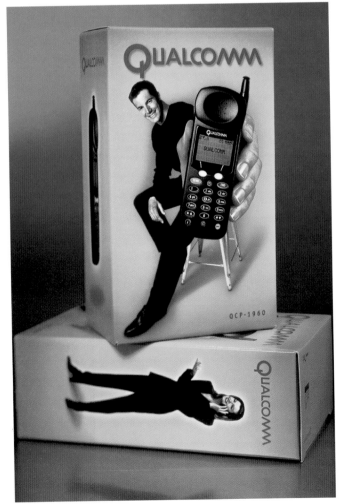

Agency: Mires Design Creative & Art Director: Jose Serrano Designer: David Adey, Deborah Hom Photographer: Carl Vanderschuit Client: Qualcomm

Agency: Mires Design Creative & Art Director: Jose Serrano Designer: Deborah Hom Photographer: Carl Vanderschuit Copywriter: Andrea May Client: Qualcomm

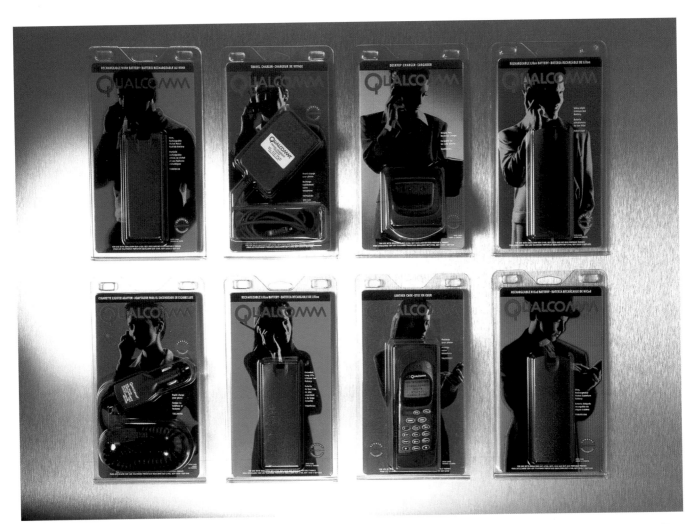

Agency: Mires Design Creative Director: Jose Serrano Designer: Deborah Hom Photographer: Carl Vanderschuit Copywriter: Andrea May Client: Qualcomm

NUTRITIVE HYDRATING

ALGAE TONER

No
5

NOURISHE

e 100 ml

VITAMIN⊕

STIMULATING

MASSAGE
AND
BATH
ORGANIC
OIL
WITH
VITAMIN

E

AND
MOUNTAIN
HERBS

50 ml | 1.7 FL.

Agency: Harcus Design Art Director: Annette Harcus Designer: Marie Milostic, Annette Harcus Photographer: Michael Rau Client: T

Agency: L3 Creative Creative Director: Lisa Ledgerwood, Mark Ledgerwood Designer: Mark Ledgerwood Photographer: Scott Baxter Client: Dial

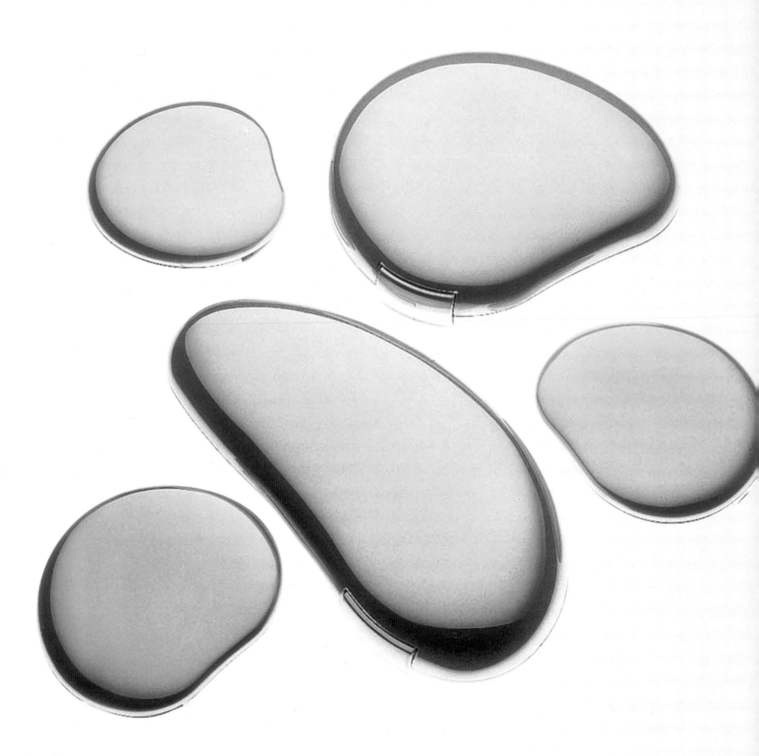

Agency: Shiseido Creation D.V.S. Creative Director: Shyuich Ikeda Art Director: Aoshi Kudo Designer: Aoshi Kudo, Izumi Matsumoto, Taisuke Kikuch, Junko Ikegaya, Keiko Hirano Client: Shiseido Co. Ltd.

BODY
POWDER

ROLL ON
ANTI-PERSPIRANT

B

Vitamin

BODY
LOTION

Vitamin

and ALOE VERA

25 ml | 8·45 fl

BODY
CREAM

A

Vitamin

and Australian Cactus Extract

200 ml | 6.75 fl oz

Agency: Harcus Design Art Director: Annette Harcus Designer: Marie Milostic, Annette Harcus Illustrator: Simon Fenton Client: Trelivings

Agency: Harcus Design Art Director: Annette Harcus Designer: Melonie Ryan Photographer: Keith Arnold Illustrator: Paul Newton Client: Trelivings

Agency: Design Ahead Designer: Theo Decker Client: Highway

Agency: Design Guys Art Director: Steven Sikora Designer: Anne Peterson, Steven Sikora Illustrator: Gary Patch Client: Bath & Body Works

modern organic products

mop basil mint shampoo

for normal to oily hair

shampooing au basilic et à la menthe cheveux normaux ou gras

If your hair feels like it's carrying the weight of the world, lift away dirt, pollutants & oil with this cleansing shampoo containing extracts of Certified Organic Peppermint, Basil, Sage & Rosemary. Leaves hair fresh & lively, full of healthy shine & bounce. Lighten up.

10.15 fl oz / 300 mL

modern organic products

mop extreme protein

treatment for damaged hair

protéine extrême crème de soin pour cheveux abîmés

Stressed out strands? Your body needs protein to stay strong and so does your hair. If your hair is weak & breaking, feed it a protein-rich diet of Wheat, Vegetable & Milk Protein as well as Certified Organic Black Beans & Olive Oil. We call it stress management.

6.76 fl oz / 200 mL

modern organic products

mop form foaming gel

light hold

gel moussant formant fixation légère

Did you hear the story about the person who wanted perfect hair? Well, Form Foaming Gel for light hold, with extracts of Certified Organic Hyssop, Nettles & Damiana, won't give you perfect hair, but it will give you good hair. And who doesn't look forward to a good hair day?

6.76 fl oz / 200 mL

modern organic products

mop styling
tonic

conditioning
fixative spray

tonic coiffant
spray coiffant
nourrissant

It's a leave-in condi-
tioner. It's a super
styler. It's a shine
builder. It's amazingly
easy to use. With
extracts of Certified
Organic Horsetail,
Dandelion, Fenugreek
& Echinacea, it helps
you do it all in a spritz.
Never leaves your hair
looking stiff or like you
tried too hard.

8.45 fl oz / 250 mL

n organic products

Want control
without stiffness?
Hold that won't feel
brittle or flake?
Defining Cream with
Certified Organic
Olive Oil & extract
of Certified Organic
Goldenseal will
leave you applaud-
ing the texture,
definition & control
that lets you be
creative. Awarded
best direction

4.23 fl oz /125 mL

modern organic products

mop glisten
shine drops

brillance
gouttes
brillance

Extracts of
Certified Organic
Oatstraw, Horsetail
& Aloe are three
shining examples
of how to add
dazzling reflection
& silkiness to
otherwise low lus-
ter hair. Contains
Vitamin E & UV
protectors too.
Great stuff.

1.7 fl oz / 50 mL

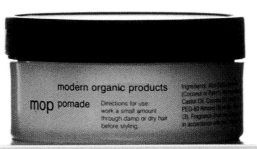

modern organic products

mop pomade

Directions for use:
work a small amount
through damp or dry hair
before styling.

Agency: Shiseido Creation D.V.S. Creative Director: Shyuich Ikeda Art Director: Aoshi Kudo Designer: Aoshi Kudo, Izumi Matsumoto, Taisuke Kikuch, Junko Ikegaya, Keiko Hirano Client: Shis
Co., Ltd.

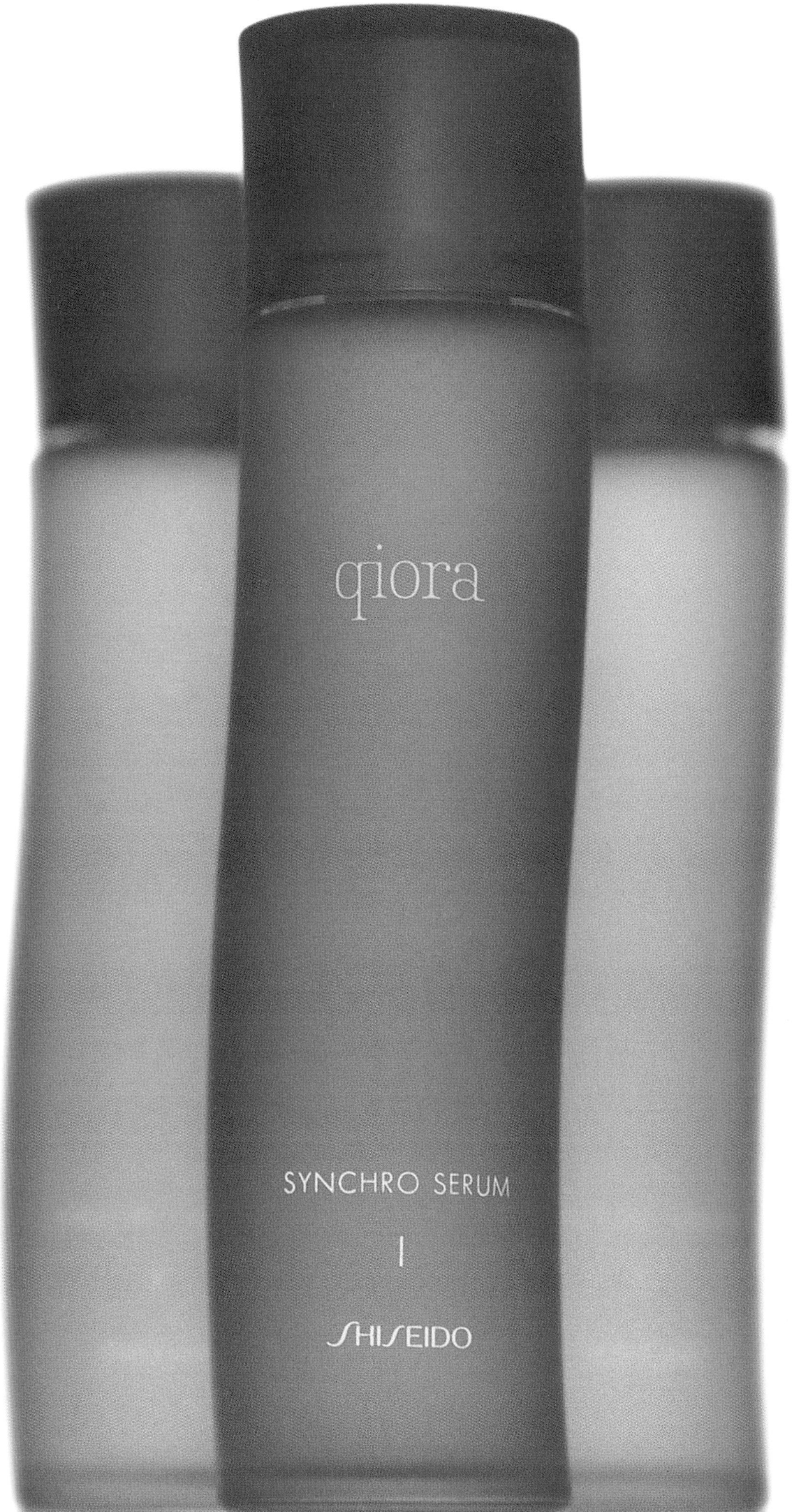
qiora
SYNCHRO SERUM
I
SHISEIDO

Agency: David Carter Design Associates Creative Director: Lori B. Wilson Art Director & Designer: Sharon LeJeune Photographer: Michael Wilson Client: Atlantis, Paradise Island

Agency: Fossil, Inc. Creative Director: Tim Hale Art Director, Designer, Illustrator & Copywriter: John Dorcas Photographer: Russ Aman Client: Fossil

Agency: Fossil, Inc. Creative & Art Director: Tim Hale Designer, Illustrator & Copywriter: Steven Zhang Photographer: Russ Aman Client: Fossil

(top) Agency: Fossil, Inc. Creative & Art Director: Steven Zhang Designer & Illustrator: Ellen Tanner Photographer: Shelley Kaler Client: Fossil (bottom) Agency: Fossil, Inc. Creative Dire
Tim Hale Art Director: Steven Zhang Designer, Illustrator & Copywriter: John Kirk Photographer: Russ Aman Client: Fossil

Agency: Fossil, Inc. Creative & Art Director: Tim Hale Designer, Illustrator & Copywriter: Steven Zhang Photographer: Russ Aman Client: Fossil (bottom) Agency: Fossil, Inc. Creative & Art tor: Steven Zhang Designer & Illustrator: Ellen Tanner, Patrick Reeves Photographer: Shelley Kaler Client: Fossil

Agency: Ceradini Design, Inc. Creative Director & Designer: David Ceradini Illustrator: Roger Sainz Client: The Great Atlantic and Pacific Tea Co.

Agency: Greteman Group Creative Director: Sonia Greteman Art Director: Sonia Greteman, James Strange Designer: James Strange Client: Chica Bella

Agency: Amway Creative Packaging Group Creative Director: Michael Horrigan Art Director: Paul Jackman Designer: Donna Smith Client: Amway Marketing/North America

Agency: Ciesa & Associates Creative Director: Michael Sundermann Art Director: Lauren Ciesa Designer: Chris Vanwyck Client: Fireside Coffee Co.

Agency: SBG Enterprise Creative Art &Director: Mark Bergman Designer: Carrie Binney, Philip Ting Photographer: Randall Ingalls Client: Peet's Coffee & Tea

Agency: Primo Angeli, Inc. Creative Director: Carlo Pagoda Art Director: Jennifer Bethke Designer: Toby Sudduth Photographer: Michael Erdman Client: R. Torre and Co., Inc.

Agency: Nestor-Stermole VCG **Creative Director:** Okey Nestor, Rick Stermole **Art Director:** Okey Nestor **Designer:** Nicole Michels, Anthony DeMarino **Illustrator:** Nicole Michels **Client:** C[ ]
Antica

cy: Leslie Evans Design Associates Creative Director: Leslie Evans, Tom Hubbard Art Director: Leslie Evans Designer: Tom Hubbard, Shoshannah White Illustrator: Bruce Hutchison
: L.L. Bean

Agency: Cornerstone Creative & Art Director: Keith Steimel Client: H.J. Heinz

Agency: Coleman Group Creative Director: Karen Corell Art Director: Joan Nicosia Designer: Joe Cuttione, Christine Kulisek Illustrator: Mike Wepplo Client: Seneca Foods, Inc.

(top) Agency: Mark Oliver, Inc. Creative & Art Director, Copywriter: Mark Oliver Designer: Patty Driskel, Mark Oliver Illustrator: Steve Salerno Client: Patti's Pickledilly Pickles (bottom) Ag
Hornall Anderson Design Works, Inc. Art Director: Jana Nishi Designer: Jana Nishi, Sonja Max Illustrator: Sticky Fingers Bakery Client: Sticky Fingers Bakery

Agency: Compass Design Designer: Mitchell Lindgren, Tom Arthur, Rich McGowan Illustrator: Cindy Lindgren Client: International Foods, Inc.

Agency: Louise Fili Ltd. **Creative & Art Director:** Louise Fili **Designer:** Louise Fili, Mary Jane Callister **Client:** California Grapeseed Oil

Agency: Auston Design Group Creative & Art Director: Tony Auston Designer: Michelle Mecchi Photographer: David Bishop Illustrator: David Laverty Client: Sparrow Lane Vinegar

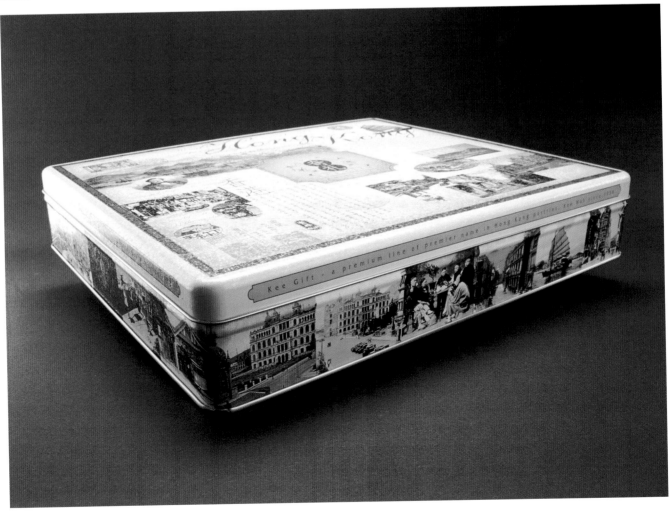

Agency: Alan Chan Design Company Creative & Art Director: Alan Chan Designer: Alan Chan, Peter Lo Client: Kee Gift Ltd., Hong Kong

Agency: Kilmer & Kilmer Creative Director: Richard Kilmer Art Director & Illustrator: Randall Marshall Designer: Randall Marshall, Gary Kohlman Client: Bite Size Bakery

Agency: Ceradini Design, Inc. Creative Director: David Ceradini Designer: Peter Novello Photographer: Ekasit Vichitlakakarn Client: The Great Atlantic and Pacific Tea Co.

Wilder Lachs - Natur pur.

Keta Salmon ist ein wilder Lachs, gelangen von der Fischerfamilie O'Connor aus den eiskalten Gewässern Alaskas. Die Familie O'Connor fängt seit Generationen nur eine kleine Menge wilder Lachse in den gleichen Fanggebieten, um deren Bestand nicht zu gefährden. Der wilde Lachs schwimmt sein Leben lang im offenen Meer und legt dabei tausende Kilometer zurück. Er ist "durchtrainiert", daher sehr fettarm und ernährt sich wie ein Gourmet von Shrimps, Garnelen, Krebsen und kleinen Fischen. Nur im wilden Lachs steckt somit eine gesunde Mischung aus Vitaminen A, B, D und E, hochungesättigte Fettsäuren, DHA Fettsäuren, Niacin, Riboflavin, Eisen, Zink, Phosphor und Magnesium. Sein hoher Anteil an Omega 3 Fettsäuren trägt bekanntlich zur Senkung des Cholesterinspiegels bei.

Der Wilde Lachs - Naturbursche und Weltenbummler - ist eine feinschmeckende Delikatesse, zubereitet nach Rezept oder frei nach Ihrer Phantasie. Seine Qualität und Feinheit benötigt keine großartigen Kochkünste - nur die Liebe zum Kochen und zum Genuß.

KETA SALMON
☆ Wilder Youkon Lachs aus Alaska ☆

Lachs im Ganzen ohne Kopf.

☆ Wilder Youkon Lachs aus Alaska ☆

◀ KETA SALMON ▶
WILDER LACHS
»NATUR PUR«

WILDER LACHS

Lachs im Ganzen ohne Kopf.

Client: Youkon Wilder Lachs

Agency: Mark Oliver, Inc. Creative Director & Copywriter: Mark Oliver Art Director: Patty Driskel Designer: Mark Oliver, Patty Driskel, Brenna Pierce Illustrator: John Lawrence Client: Ocea Beauty Seafood

cy: The Weightman Group Creative Director: Nat Gutwirth Art Director & Designer Tom Loder Photographer: John Romeo Illustrator: Jim Starr Copywriter: Catherine Wetendorf Client:
iy's Microbatch Ice Cream

Agency: Pentagram Design Creative & Art Director: Kit Hinrichs Designer: Brian Jacobs Client: Dreyers Grand Ice Cream

Agency: Bread & Butter Art Director: Kelly Hensley Photographer: Kathryn MacDonald Copywriter: Adolfo Calero Client: Lettieri & Co.

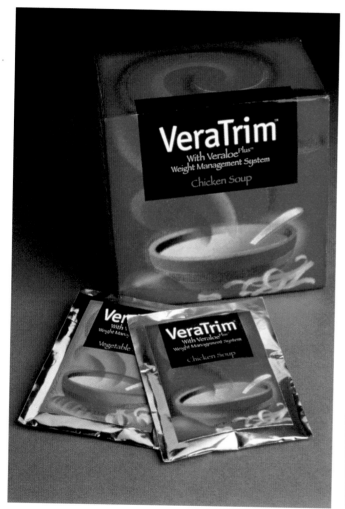

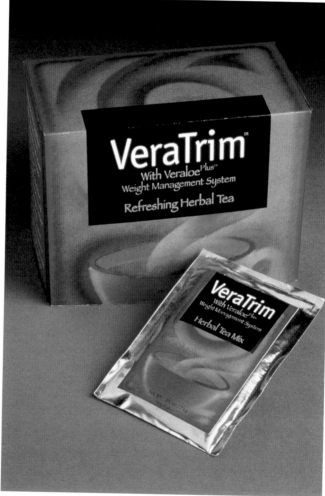

Agency: Sibley Peteet Design Creative Director: Don Sibley Art Director, Designer & Illustrator: Tom Hough Client: BeautiControl Cosmetics

Agency: Louise Fili, Ltd. Creative & Art Director: Louise Fili Designer: Louise Fili, Mary Jane Callister Client: Bella Cucina

Agency: Louise Fili. Ltd. Creative & Art Director: Louise Fili Designer: Louise Fili, Mary Jane Callister Client: Jean-Georges Vongrichten

Agency: Louise Fili Ltd. Creative & Art Director: Louise Fili Designer: Louise Fili, Mary Jane Callister Illustrator: James Grashen Client: El Paso Chile Co.

Agency: Louise Fili Ltd. Creative & Art Director: Louise Fili Designer: Louise Fili, Mary Jane Callister Client: Bella Cucina

Agency: Art Force Studio Art Director & Designer: Halasi Zoltan Client: Herbaria Rt.

Agency: Acme Graphic Design Art Director & Designer: Greg Mally Illustrator: Anton Kimball Client: Stash Tea Company

Agency: Robert Bailey Inc. Creative Director & Art Director, Designer, Copywriter: Dan Franklin Photographer: Peter Rose Client: Boyd Coffee Company

(top) Agency: Learning Resources **Creative Director:** Diane Jacobs **Designer:** Diane Jacobs, Tirza Ernst **Illustrator:** Patrick Merrell **Client:** Leaning Resources **(bottom) Agency:** Learı Resources **Creative Director:** Diane Jacobs **Designer:** Diane Jacobs, Tirza Ernst **Illustrator:** Mike Takagi **Client:** Learning Resources

acy: Hornall Anderson Design Works, Inc. Art Director: Jack Anderson, Jana Nishi Designer: Jana Nishi, Sonja Max Illustrator: Denise Weir Copywriter: Resource Games Client: Resource es

Agency: Greteman Group **Creative & Art Director:** Sonia Greteman **Designer:** James Strange **Client:** Envision

Agency: Design Guys Creative Director: Steven Sikora Photographer: Amy Kirkpatrick, Jim Erickson Copywriter: Cynthia Miller Client: Target Stores

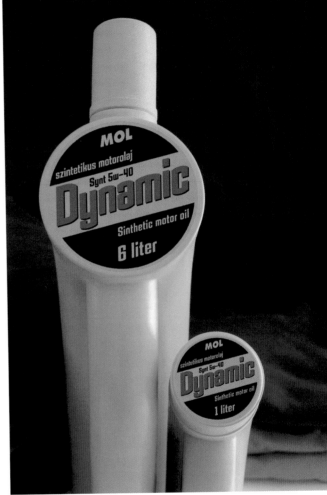

Agency: Art Force Studio Art Director: Simon Attila Designer: Simon Attila, Toth Judit Client: Mol Hungarian Oil & Gas Company

Agency: Turner Duckworth Creative Director: David Turner, Bruce Duckworth Art Director: Janice Davison Designer: Mark Waters Client: Superdrug

Agency: Design North, Inc. Creative Director: Gwen Granzow Designer: Gwen Granzow, Jackie Lagenecker Client: Kaytee Products, Inc.

Agency: MLR Design Creative Director: Linda Voll Art Director: Linda Voll Designer: Linda Voll Illustrator: Leslie Wu Client: Gutwein

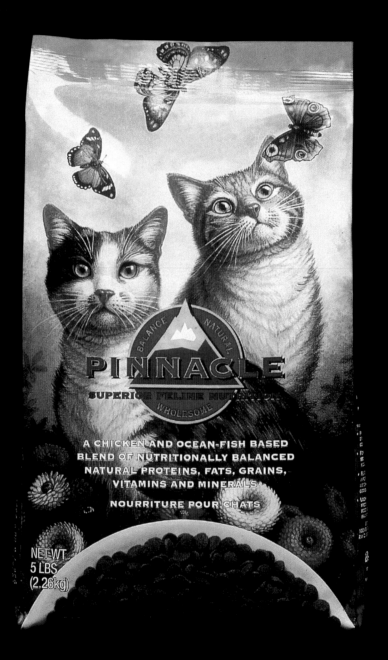

Agency: Addis Creative Director: Joanne Hom Designer: David Leong Illustrator: Hamagami Carroll & Associates Client: Ortho

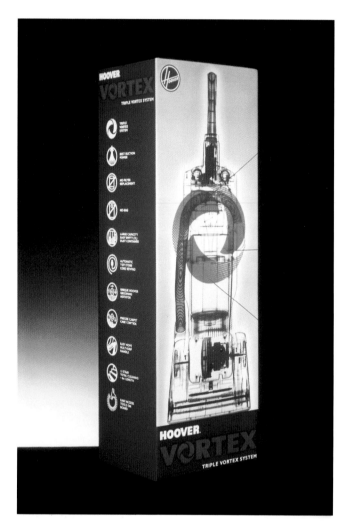

Agency: Coley Porter Bell Creative & Art Director: Allison Miguel Designer: Victoria Fletcher Photographer: Nick Veassey Client: Hoover

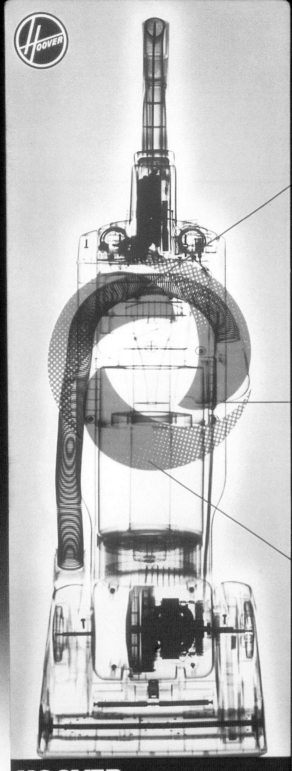

Agency: L3 Creative Creative Director: Lisa Ledgerwood, Mark Ledgerwood Art Director: Lisa Ledgerwood Designer: Julie Scott Client: Belae Brands

Agency: Desgrippes Gobé Creative Director: Phyllis Aragaki Art Director: Nathalie Jacobs Client: Leap Energy & Power Corp.

Agency: Amway Creative Packaging Group **Creative Director:** Michael Horrigan **Art Director & Designer:** Paul Jackman **Client:** Amway Marketing/North America

maharam

Panel Fabric

Systems &
Upholstered Walls 1

maharam

Cubicle, Drapery, &
Bedspread Fabrics 1

maharam

Textiles & Wallcovering

Fabrics & Papers 1

maharam

Upholstery

Patterns 4

maharam

maharam

Vinyl Wallcovering

Solids & Textures 1

Agency: Plus One Design Creative Director: Amulya Baruah Client: Indian Oil Corporation

Agency: Hornall Anderson Design Works, Inc. Creative & Art Director: Jack Anderson, Lisa Cerveny Designer: Jack Anderson, Lisa Cerveny, Alan Florsheim, David Bates, Mike C
Copywriter: Leatherman Tool Group Client: Leatherman Tool Group

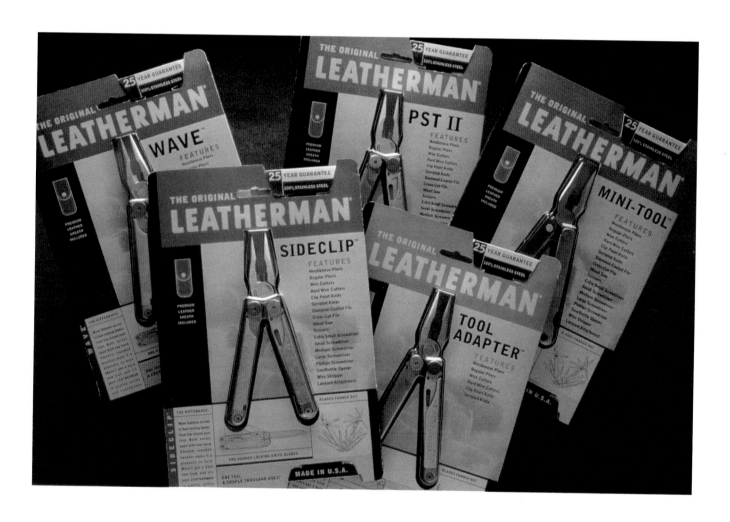

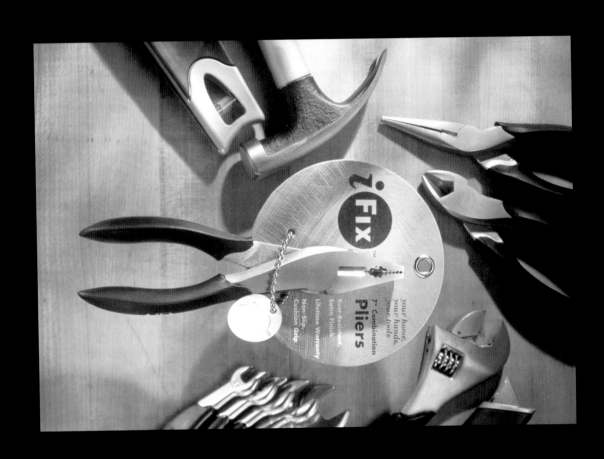

Agency: Larsen Design & Interactive Creative Director: Paul Wharton Designer: Peter de Sibour, Chris Zastoupil, Todd Mannes Copywriter: Pam Powell Client: Target Stores

Agency: Mires Design Creative & Art Director: Jose Serrano Designer: Miguel Perez Illustrator: Tracy Sabin, Nancy Stahl Client: Deleo Clay Tile Company

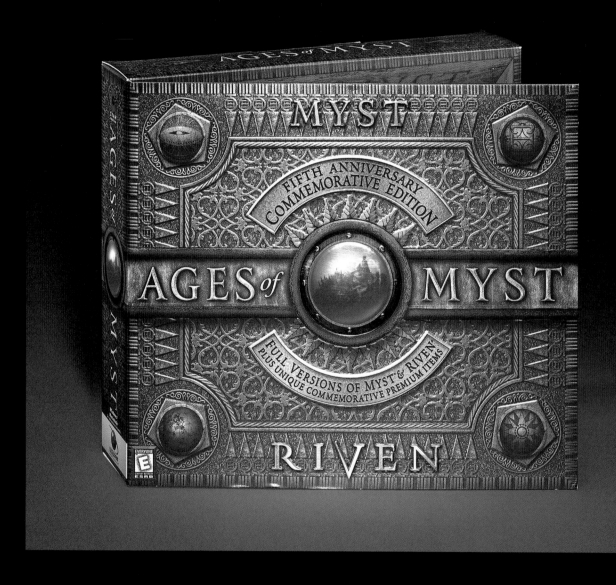

Agency: Deutsch Design Works Creative Director: Barry Deutsch Designer: Lori Wynn, Jaques Rossouw, Eric Pino, John Lucas Client: Red Ord Div. of Broderbund

Agency: Mires Design Creative & Art Director: Jose Serrano Designer: Jeff Samaripa Photographer: David Deahl Illustrator: Miguel Perez Client: Big Deahl, Inc.

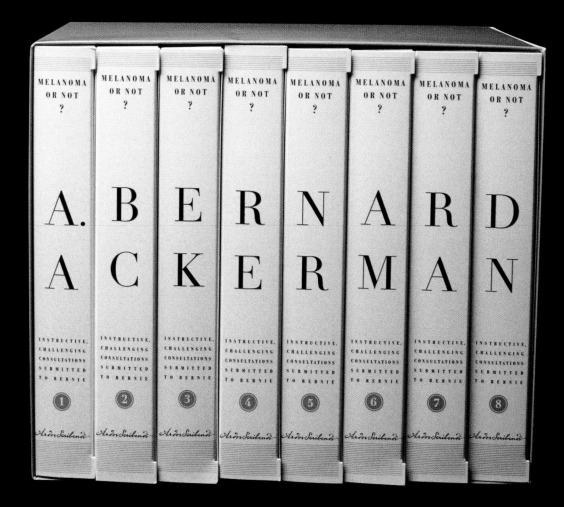

Agency: Louise Fili, Ltd. Creative & Art Director: Louise Fili Designer: Louise Fili, Mary Jane Callister Client: Dr. A. Bernard Ackerman

Agency: Duffy Design Creative Director: Joe Duffy Art Director, Designer& Illustrator: Tom Riddle Copywriter: John Jarvis Client: International Navistar

cy: Hornall Anderson Design Works, Inc. Creative Director: Larry Anderson Art Director: John Hornall, Katha Dalton Designer: Katha Dalton, Larry Anderson, Alan Florshiem, Bruce Stigler, Farmer Client: Microsoft Corporation

Agency: Michael Osborne Design Creative Director: Michael Osborne Designer: Michelle Regenbogen Photographer: David Wakely Client: SFMOMA Museum Store

y: Sagmeister Inc. Creative & Art Director: Stefan Sagmeister Designer: Hjalti Karlsson Photographer: Susan Stava Illustrator: Barbara Ehrbar Copywriter: Jamie Block Client: Capitol
ds

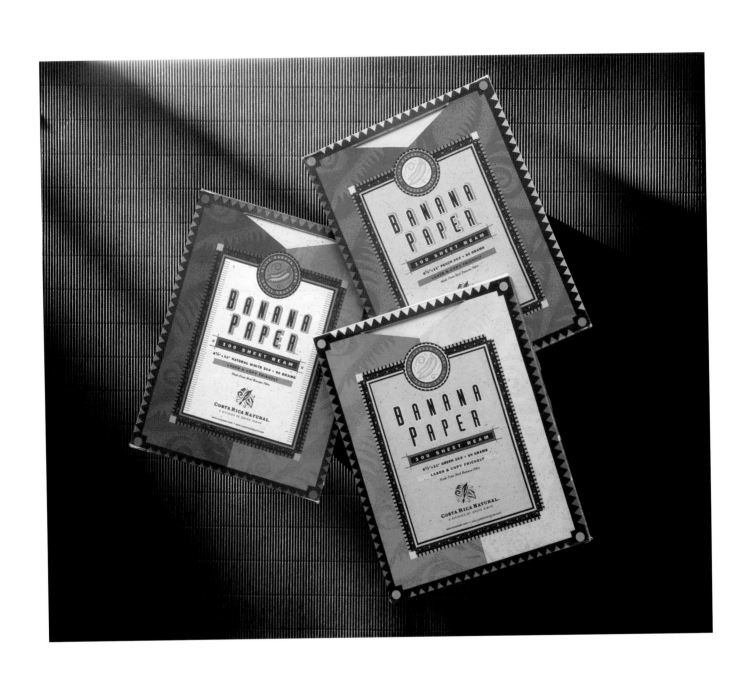

Agency: Mires Design Creative & Art Director: Jose Serrano Designer: Miguel Perez Illustrator: Tracy Sabin Client: Green Field Paper Company

Agency: Mires Design Creative & Art Director: Jose Serrano Designer: Miguel Perez Illustrator: Tracy Sabin Client: Green Field Paper Company

Agency: Greteman Group Art Director: Sonia Greteman, James Strange Designer: James Strange Client: Costa Rica Natural Paper

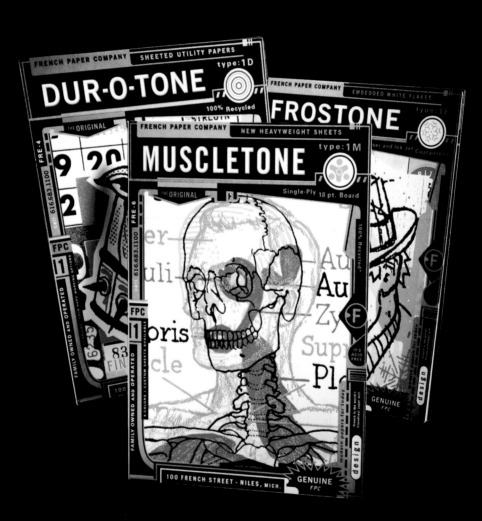

cy: Charles S. Anderson Design Art Director: Charles S. Anderson Designer: Todd Piper-Hauswirth, Kyle Hames Illustrator: Charles S. Anderson, Kyle Hames Copywriter: Lisa Pemrick
: French Paper Company

Agency: Kilmer & Kilmer Art Director: Richard Kilmer Designer: Randall Marshall Client: New Leaf, Inc.

Agency: Turner Duckworth Creative Director: David Turner, Bruce Duckworth Designer: Bruce Duckworth Illustrator: Justin Delavison Client: Superdrug

STIMULATING BATH OIL
A DISPERSING BATH OIL TO RESTORE VITALITY

BENZOIN, GRAPESEED (VITIS VINIFERA) OIL, GLYCERIN, POLYSORBATE-60, LAVENDER
(LAVANDULA OFFICINALIS) OIL, WHEATGERM (TRITICUM VULGARE) OIL, ALMOND (PRUNUS
DULCIS) OIL, CARROT (DAUCUS CAROTA) OIL, JASMINE (JASMINUM OFFICINALIS) OIL, SPEARMINT
(MENTHA VIRIDIS) OIL, GERANIUM (PELARGONIUM GRAVEOLENS) OIL, BENZOIN (STYRAX BENZOIN) OIL,
METHYLPARABEN, ETHYLPARABEN, PROPYLPARABEN, BUTYLPARABEN, DISODIUM EDTA.

MADE IN ENGLAND

CALENDULA SHAMPOO
FOR DRY SKIN

A NON-GREASY CLEANSING LOTION CONTAINING CLEANSING AND STRINGENT HERBS
WIPE OVER THE SKIN USING COTTON WOOL AND RINSE OFF WITH LUKEWARM WATER.
INGREDIENTS: WATER (AQUA), ELDERFLOWER (SAMBUCUS NIGRA) EXTRACT, ALOE VERA (ALOE BARBADENSIS)
GRAPESEED (VITIS VINIFERA) OIL, STEARIC ACID, APRICOT KERNEL (PRUNUS ARMENIACA) OIL, ALMOND (PRUNUS
DULCIS) OIL, CETYL STEARYL ALCOHOL, WITCH HAZEL (HAMAMELIS) EXTRACT, SAGE (SALVIA OFFICINALIS)
SAGE (SALVIA OFFICINALIS) EXTRACT, BEESWAX, TRIETHANOLAMINE, POLYSORBATE, METHYL
PARABEN, COCONUT(COCOS NUCIFERA) OIL, SHEA NUT (BUTYROSPERMUM PARKII) BUTTER,
LEMONGRASS(CITRUS LIMONUM) OILSAGE,(SALVIA OFFICINALIS) OIL LAVENDER (LAVENDULA OFFICINALIS)

LEMONGRASS SUN LOTION
FOR OILY AND PROBLEM SKIN

A LIGHT, NON-GREASY CLEANSING LOTION CONTAINING CLEANSING AND STRINGENT
EXTRACTS, WIPE OVER THE SKIN USING COTTON WOOL AND RINSE OFF WITH
INGREDIENTS: WATER (AQUA), ELDERFLOWER (SAMBUCUS NIGRA) EXTRACT, ALOE VERA
GRAPESEED (VITIS VINIFERA) OIL, STEARIC ACID, APRICOT KERNEL (PRUNUS ARMENIACA)
DULCIS) OIL, CETYL STEARYL ALCOHOL, WITCH HAZEL (HAMAMELIS) EXTRACT, SAGE (SALVIA
OFFICINALIS) EXTRACT, LAVENDER(LAVENDULA)(OFFICINALIS) EXTRACT, BEESWAX, TRIETHANOLAMINE, POLYSORBATE
METHYL PARABEN, COCONUT(COCOS NUCIFERA) OIL, SHEA NUT (BUTYROSPERMUM PARKII) B
LEMONGRASS(CITRUS LIMONUM) OILSAGE,(SALVIA OFFICINALIS) OIL LAVENDER (LAVENDULA

MADE IN ENGLAND

Agency: Pentagram Design Creative & Art Director: Kit Hinrichs Designer: Brian Jacobs Client: Potlatch

Agency: Compass Design Designer: Mitchell Lindgren, Tom Arthur, Rich McGowan, Sharon Sudman Client: Compass Design

Agency: Graphic Solutions, Inc. Creative & Art Director: Marc Tebon Designer: Steve Radtke Client: Graphic Solutions, Inc.

Agency: Harcus Design Art Director & Designer: Annette Harcus Client: Southern Star

Agency: Watts Design Creative & Art Director, Designer: Peter Watts Client: Watts Design

Agency: Mires Design Creative & Art Director: Jose Serrano Designer: Miguel Perez Illustrator: Tracy Sabin Client: Bordeaux Printers

Agency: Wallace Church Associates, Inc. Creative & Art Director, Designer: Nin Glaister Client: Wallace Church Associates, Inc.

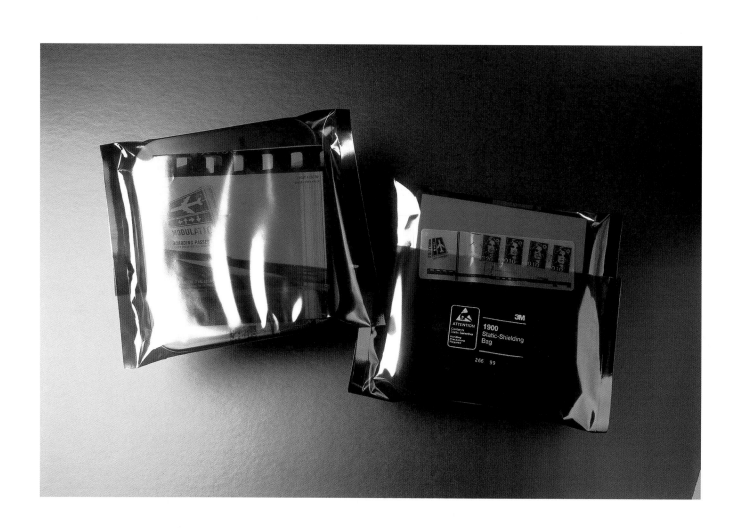

Agency: Iridium Marketing & Design Creative Director: Etienne Bessett, Mario L'Écuyer Art Director: Jean-Luc Denat Designer & Illustrator: Etienne Bessette Photographer: Iridium Marke
& Design Copywriter: Stephen J. Hards Client: Iridium Marketing & Design

y: Iridium Marketing & Design Creative Director: Lucero Sánchez Art Director: Jean-Luc Denat Designer: Etienne Bessette Photographer: Headlight Innovative Imagery Copywriter:
en J. Hards Client: Iridium Marketing & Design

Agency: The Traver Company Client: The Traver Company

Agency: Shin Matsunaga Design, Inc. Creative & Art Director, Designer: Shin Matsunaga Client: Issey Miyake

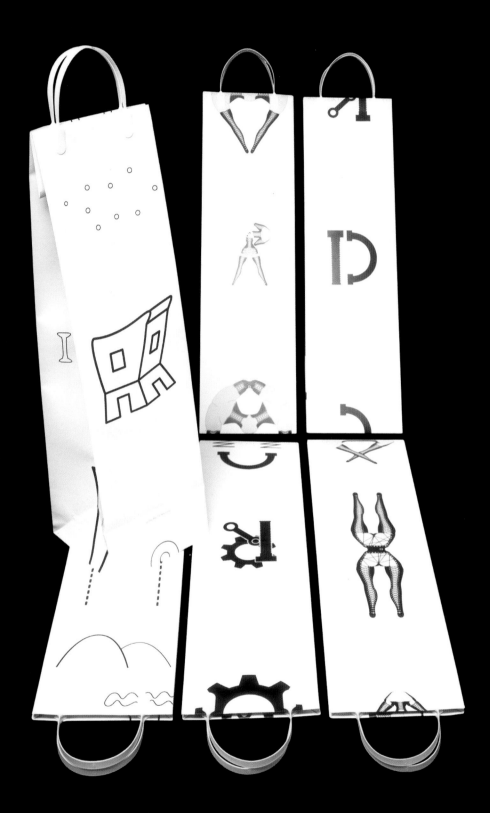

Agency: Morgane Le Fay Creative & Art Director, Designer: Bernice Pan Illustrator: Julie Kim Client: Morgane Le Fay

Agency: Greteman Group Creative Director: Sonia Greteman Art Director: Sonia Greteman, James Strange Designer: James Strange Illustrator: James Strange Client: Costa Rica Natural

y: Greteman Group Creative Director: Sonia Greteman Art Director: Sonia Greteman, James Strange Designer& Illustrator: James Strange Copywriter: Deanna Harms Client: Costa Rica
al Paper

Agency: Cornerstone Creative & Art Director, Designer: Keith Steimel Client: Estee Lauder

Agency: Cochran & Associates Creative & Art Director, Designer, Illustrator: Bobbye Cochran Client: Endo-Exo Apothecary

Agency: Package Land Co., Ltd. Creative & Art Director, Designer: Yasuo Tanaka Client: Package Land Co., Ltd.

Shopping Bags 228, 229

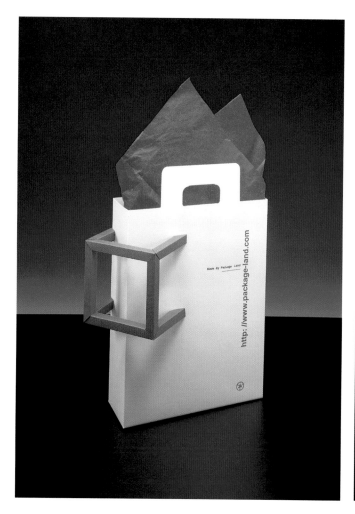
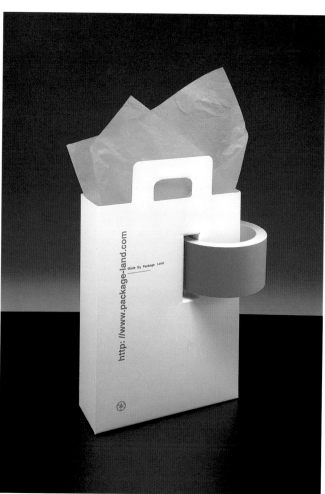

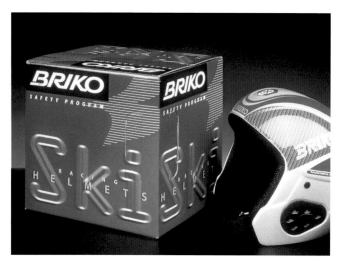

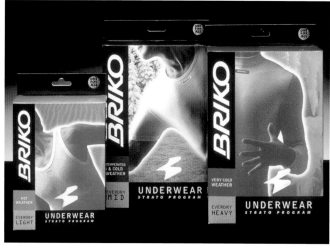

(top, left) Agency: Tangram Strategic Design Creative Director: Enrico Sempi Art Director & Designer: Antonella Trevisan Client: Briko (top, right) Agency: Tangram Strategic Design Cre
Director: Enrico Sempi Art Director: Antonella Trevisan Designer: Antonella Trevisan, Anna Grimaldi Client: Briko (middle, left) Agency: Tangram Strategic Design Creative Director: E
Sempi Art Director& Designer: Antonella Trevisan Client: Briko (middle, right) Agency: Tangram Strategic Design Creative Director: Enrico Sempi Art Director: Antonella Trevisan Desi
Antonella Trevisan, Anna Grimaldi Client: Briko (bottom, left) Agency: Tangram Strategic Design Creative Director: Enrico Sempi Art Director: Antonella Trevisan Designer: Antonella Trev
Anna Grimaldi Client: Briko (bottom, right) Agency: Tangram Strategic Design Creative & Art Director: Enrico Sempi Designer: Enrico Sempi, Antonella Trevisan, Anna Grimaldi Photogra
Edoardo Mari Illustrator: Sergio Quaranta Client: Briko

Agency: Tangram Strategic Design Creative Director: Enrico Sempi Art Director & Designer: Antonella Trevisan Client: Briko

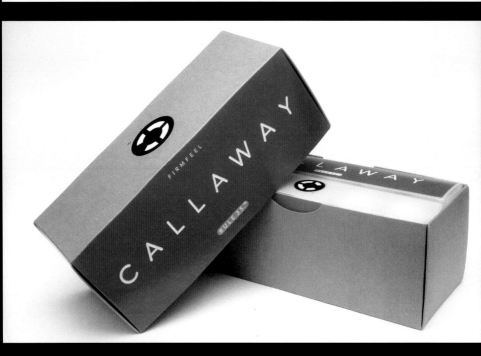

SRAM
ESP®
9.0SL
REAR
DERAILLEUR

ESP is a mountain bike specific
system design with a unique 1:1
Actuation Ratio that results in
minimal shifter effort and optimum
fast lever shifts. This 1-speed
derailleur is rated for competition
level performance.

9.0SL

7.0

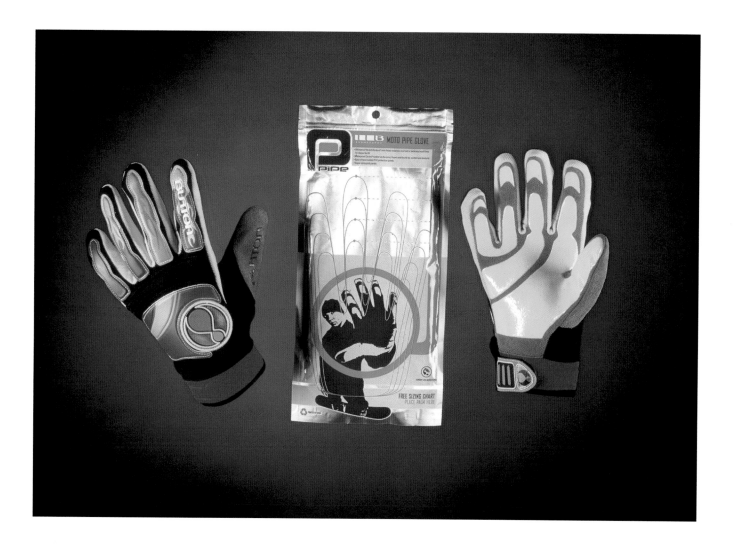

Agency: Jager Di Paola Kemp Design Creative Director: Michael Jager Art Director: Jim Anfuso Designer: Jared Eberhardt Client: Burton Snow Board

Agency: Greteman Group **Creative & Art Director:** Sonia Greteman **Designer & Illustrator:** James Strange, Sonia Greteman **Copywriter:** Deanna Harms, Sonia Greteman **Client:** Gran Colun

Agency: Deutsch Design Works Creative Director: Barry Deutsch Designer: Lori Wynn, Jacques Rossouw Client: Napa Cigar Company

Agency: Design Factory International Creative Director: Red Maxwell Designer: Grisson Davis, Jessica Stair Illustrator: Tim Anderson Client: R.J. Reynolds

Agency: Michael Stanard, Inc. Art Director: Michael Stanard Designer: Kristy Vandekerckhove Client: The Sisyphus Enterprise

CELEBRATING THE LEGENDARY TRAINS OF THE POSTWAR ERA

The year is 1945. The War is over and the nation is in a celebratory mood. And Joshua Lionel Cowen is preparing his factory to renew production of his wonderful toy trains. It will pr— the most celebrated on— legendar—

every American boy's dream and fathers helped create memories that would last a lifetime. It is in this spirit that the artisans and engineers of Lionel have crafted this commemorative series. It's genuine Lionel – the perfect symbol of the limitless optimism and potential of Postwar America.

Agency: Michael Stanard, Inc. Art Director: Michael Stanard Designer: Marc Fuhrman Client: Lionel Trains

Agency: Michael Stanard, Inc. Art Director: Michael Stanard Designer & Illustrator: Michael Chang Client: Lionel Trains

Agency: Design Ahead Creative Director: Axel Voss Designer: Theo Decker Photographer: Detlef Odenhansen Client: Gmund

Index

AgenciesClients

1881 Tabacalera 240, 241

A Design, La Mirada, CA (562) 941 2090 215
Ackerman, Dr. A. Bernard 184
Acme Graphic Design, Portland, OR (503) 872 8940 . . . . 67, 154
Addis, Berkeley, CA (510) 704 7500 . . . . 144, 167
Agassi Enterprises 234
Alan Chan Design Company, Hong Kong, China (852) 2527 8228 134
American Oak Preserving Company 168
Amway Creative Packaging Group, Ada, MI (616) 787 5168 . . . . . . 122, 174
Amway Marketing/North America 122, 174
Anheuser-Busch 20, 23
Annegret Beier, Paris, France 33 1 453 33377 . . . . 102, 103
Apothecary Tinctura 200
Arcus Produkter AS 52
Arias Associates, Palo Alto, CA (650) 321 8138 . . . . 240, 241
Arietta 62
Art Force Studio, Budapest, Hungary, 36 1339 5348 . . . . 153, 160, 169
Atlantis, Paradise Island 110
August Schells Brewing 18
Auston Design Group, Saint Helena, CA (707) 963 4152 . . . . 62, 133

Bath & Body Works 105
BeautiControl Cosmetics 147
Belae Brands 172
Bella Cucina 148, 149, 152
Big Deahl, Inc. 183
Bill Carson Design, Austin, TX (512) 371 1150 39
Bite Size Bakery 135
Blackburn's Ltd., London, England, 171 734 7646 . . . .
. . . . 32, 33, 43, 56, 57, 69, 75, 81
Bordeaux Printers 213
Boyd Coffee Company 155
Branion GmbH–On Top 186
Bread & Butter, San Francisco, CA (415) 282 8678 145
Breeders Choice Pet Food 164, 165
Briko 230, 231
Britton Design, Sonoma, CA (707) 938 8378 . . . . 64, 65, 66
Buena Vista Winery 66
Burton Snow Board 235

Caldewey Design, Napa, CA (707) 252 6666 . . . . 35, 36, 37, 74
California Grapeseed Oil 132
Callaway 232
Capitol Records 191
Ceradini Design, Inc., New York, NY (212) 255 7277 . . . . 55, 118, 136
Charles S. Anderson Design, Minneapolis, MN (612) 339 5181 . . . . . 196, 197
Chica Bella 120
Ciesa & Associates, East Lansing, MI (517) 351 2453 123
Clarity Coverdale Fury 22
Cochran & Associates, Chicago, IL (773) 404 0375 227
Coleman Group, New York, NY (212) 375 0550 129
Compass Design, Minneapolis, MN (612) 339 1595 . . . . 18, 131, 208
Coley Porter Bell, London, England 171 470 4000 . . . . 170, 171
Cornerstone, New York, NY (212) 686 6046 . . . . 27, 128, 226, 238, 239
Costa Rica Natural Paper 192, 195, 224, 225
Covent Garden Soup Co. 146
Cucina Antica 127

David Carter Design Associates, Dallas, TX (214) 826 4631 110
Deleo Clay Tile Company 181
Desgrippes Gobé, New York, NY (212) 979 8900 173

Design Ahead, Essen, Germany 49 201 842060 . . . . 101, 186, 248
Design Factory International, Winston Salem, NC (336) 760 0770 . . 242, 243
Design Guys, Minneapolis, MN (612) 338 4462 . . . . 105, 159
Design North, Inc., Racine, WI (262) 639 2080 162
Design Ranch, Kansas City, MO (816) 472 8668 200
Deutsch Design Works, San Francisco, CA (415) 487 8520 . . . .
. . . . 20, 21, 23, 72, 146, 182, 237
Deutsch Inc. (212) 981 7600 . . . . 2, 50, 51
Dial 93
Dreyers Grand Ice Cream 142, 143
Duck Pond Cellars 67
Duffy Design, Minneapolis, MN (612) 321 2333 185

El Paso Chile Co. 151
Endo-Exo Apothecary 227
Envision 159
Estee Lauder 226

Fireside Coffee Co. 123
Firestone/Walker 24
Fossil, Inc. 112, 113, 114, 115
French Paper Company 196, 197

Gaetano Specialties, Ltd. 46, 48
Gambrinus Company 25
Gekeikan Inc. 42, 73
Gianna Rose 100
Gillis & Smiler, Los Angeles, CA (310) 208 5645 203
Gmund 248
Gran Columbia 236
Granny Goose 144
Graphic Solutions, Inc., Milwaukee, WI (414) 228 9666 209
Great Atlantic & Pacific Co. of Canada 41, 118, 137
Great Northern Brewing Co. 21
Green Field Paper Company 193, 194
Greteman Group, Wichita, KS (316) 263 1004 . . . .
. . . . 120, 159, 192, 195, 224, 225, 236
Guld & Grüna Skogar Design, Stockholm, Sweden 46 858 70 8000 78
Gustav Klimt Parfums 91
Gutwein 163
Gymboree 116, 117

Harcus Design, Surry Hills, Australia 61 2 9212 2755 . . 59, 60, 92, 98, 99, 210
Herbaria Rt. 153
Highway 101
H.J. Heinz 128
Honolulu Coffee Co. 121
Hoover 170, 171
Hornall Anderson Design Works, Inc., Seattle, WA (206) 467 5800 . . . .
. . . . 130, 157, 178, 179, 187

Indian Oil Corporation 176, 177
Influence Branding, St. Louis, MO (314) 588 1500 . . . . 201, 202
International Foods, Inc. 131
International Navistar 185
Ipsa Co., Ltd. 96, 97
Iridium Marketing & Design, Ottawa, Canada (613) 748 3336 . . . . . 216, 217
Issey Miyake 222

Jager Di Paola Kemp Design, Burlington, VT (802) 864 5884 235
James Sadler 166
JBB Greater Europe 44

Jean-Georges Vongrichten 150

Jennifer Sterling Design, San Francisco, CA (415) 612 3481 . . . . 90, 91

Jeremy's Microbatch Ice Cream 141

Joe Boxer Corporation, San Francisco, CA (415) 882 9406 80

Jose Cuervo 47, 49

Kaytee Products, Inc.. 162

Kee Gift, Ltd., Hong Kong 134

Kilmer & Kilmer, Albuquerque, NM (505) 232 6103 . . . . 135, 198

Klim Design, Avon, CT (860) 678 1222 . . . . 47, 49

L3 Creative, Phoenix, AZ (602) 954 6992 . . . . 93, 172

Larsen Design & Interactive, Minneapolis, MN (888) 590 4405 180

Leap Energy & Power Corp 173

Learning Resources, Vernon Hills, IL (847) 573 8400 156

Leatherman Tool Group. 178, 179

Leslie Evans Design Associates, Portland, ME (207) 874 0102 127

Lettieri & Co.. 145

Lionel Trains 246, 247

Liska & Associates, Inc., New York, NY (212) 627 3200 . . . . 106, 107

L.L. Bean 127

Lloyd Ferguson Hamlins, London, England 441 777 068 762 44

Logos Identity by Design Ltd., Toronto, Canada (416) 259 7834 . . . . 41, 137

Louise Fili, Ltd., New York, NY (212) 989 9153 . . . .
. . . . 71, 121, 132, 148, 149, 150, 151, 152

Maharam 175

Marich Candies 119

Mario Milostic Design, Sydney, Australia, 02 9929 2458 . . . . 45, 211

Mark Anthony Brands 26, 74

Mark Oliver, Inc., Santa Barbara, CA (805) 963 0734 . . . .
. . . . 24, 38, 119, 130, 140, 164, 165

Mastandrea Design Inc. (415) 538 8130. 26

Matsumoto Inc., New York, NY (212) 807 0248. 175

Matt Brothers. 71

McKenzie River Corp. 16, 19

Michael Osborne Design, San Francisco, CA (415) 255 0125 . . . .
. . . . 63, 116, 117, 188, 189

Michael Stanard, Inc., Evanston, IL (847) 869 9820 . . . . . 168, 245, 246, 247

Microsoft Corporation 187

Mires Design, San Diego, CA (619) 234 6631 . . . .
. . . . 82, 83, 84, 85, 181, 183, 193, 194, 213, 234

MLR Design, Chicago, IL (312) 943 5995 163

Modern Organic Products 106, 107

Mol Hungarian Oil & Gas Company. 160

Moonlight Tobacco Co.. . . . . 238, 239

Morgane Le Fay, New York, NY (212) 334 6151. 223

Mountain Sun Organic Juices 38, 39

Napa Cigar Company. 237

Neal's Yard. 204, 205

Nestor-Stermole VCG, New York, NY (212) 229 9377 126

New Leaf, Inc. 198

Nourishe 90, 91

Nutraceutics 201, 202

Ocean Beauty Seafood 140

Old Benchmark Company 54

Orchid Drinks 32, 33, 43, 81

Ortho 167

Pacific Sensuals. 203

Package Land Co., Inc., Osaka, Japan 816 6 6675 0118. . . . . 228, 229

Packaging Create, Inc., Osaka, Japan 816 6 6941 9618 . . . . 42, 73

Parachute Design, Minneapolis, MN (612) 359 4387 22

Parfums Cacharel Paris 102, 103

Patti's Pickledilly Pickles 130

Peet's Coffee & Tea 124

Pentagram Design, San Francisco, CA (415) 896 0499. . . . .
. . . . 76, 77, 142, 143, 206, 207, 232

Plus One Design, London, England, 0181 452 1022. . . . . 176, 177

Potlatch. 206, 207

Primo Angeli, Inc., San Francisco, CA (415) 551 1900. 125

Qualcomm. 82, 83, 84, 85

R. Torre & Company Inc.. 125

R.J. Reynolds 242, 243

Red Orb Division of Broderbund 182

Resource Games. 157

Revolution Hardrinks 35, 36, 37

Robert Bailey Inc., Portland, OR (503) 228 1381. 155

Robert Mondavi Winery. 72

Sagmeister, Inc., New York, NY (212) 647 1789 191

Sapp Design, Salzburg, Austria 43 662 840 440 . . . . 111, 138, 139

Sayles Graphic Design, Des Moines, IA (515) 279 2922. 100

Sazerac Company 53

SBG Enterprise, San Francisco, CA (415) 391 9070. 124

Scandivavian Design Group, Oslo, Norway 47 22 549500 52

Seneca Foods, Inc.. 129

SFMOMA Museum Store 188, 189

Shin Matsunaga Design, Inc., Tokyo, Japan 81 3 5353 0170 . . . 29, 30, 31, 222

Shiseido Cosmetics, New York, NY (212) 805 2385 . . . . 87, 88, 221

Shiseido Creation D.V.S., Toyko, Japan 81335725111 . . 94, 95, 96, 97, 108, 109

Sibley Peteet Design, Dallas, TX (214) 969 1050 . . . . 25, 147

Sisyphus Enterprise, The. 245

Skânmeerier 78

Sociedade Dos Vinhos Borges 75

Sociedade Quinta Do Portal SA 56, 57, 69

Southern Star 210

Spar Advertising, New Orleans, LA (504) 484 7695. . . . . 53, 54

Sparrow Lane Vinegar 133

Sram Corp.. 233

Stash Tea Company. 154

Sticky Fingers Bakery 130

Superdrug. 161, 199

Swiss Dairy 76, 77

Takara Shuzo Co., Ltd.. . . . . 29, 30, 31

Tallara Wines 60

Tangram Strategic Design, Novara, Italy 39 0321 35662. . . . . 230, 231

Tanqueray 2, 50, 51

Target Stores 159, 180

Tayburn, Edinburgh, Scotland 131 662 0662. 166

Traver Company, The, Seattle, WA (206) 441 0611 218

Trelivings. 92, 98, 99

Twisted Brewing Co.. 27

Turner Duckworth, San Francisco, CA (415) 495 8691 . . . .
. . . . 16, 19, 161, 199, 204, 205

United Distillers & Vintners North America. 55

Viansa Winery. 64, 65

Vinograd Croatia 45

Van Noy Group, Torrance, CA (310) 329 0800 . . . . 46, 48

VSA Partners, Inc., Chicago, IL (312) 42 76413 233

Wallace Church Associates, Inc., New York, NY (212) 755 2903 . . . .
. . . . 214, 219

Watts Design 212

Weightman Group, The, Philadelphia, PA 141

Wente Vineyards 63

Westel 900 169

Work Beer 17

Work Inc., Richmond, VA (804) 358 9366 17

Yalumba Winery 59   Youkon Wilder Lachs 138, 139

CreativeDirectorsArtDirectorsDesigners

Adey, David, San Diego, CA (619) 234 6631 82

Allen, Peter T., San Francisco, CA (415) 882 9406 80

Anderson, Charles S., Minneapolis, MN (612) 339 5181 . . . . 196, 197

Anderson, Jack, Seattle, WA (206) 467 5800 . . . . 157, 178, 179

Anderson, Larry, Seattle, WA (206) 467 5800 187

Anfuso, Jim, Burlington, VT (802) 864 5884 235

Aragaki, Phyllis, New York, NY (212) 979 8900 173

Arias, Mauricio, Palo Alto, CA (650) 321 8138 . . . . 240, 241

Arthur, Tom, Minneapolis, MN (612) 339 1595 . . . . 18, 131, 208

Attila, Simon, Budapest, Hungary 0036 1339 5348 160

Auston, Tony, Saint Helena, CA (707) 963 4152 . . . . 62, 133

Barrett, Susanna, New York, NY (212) 627 3200 . . . . 106, 107

Baruah, Amulya, London, England 181 452 1022 . . . . 176, 177

Bates, David, Seattle, WA (206) 467 5800 . . . . 178, 179

Beier, Annegret, Paris, France 331 453 3377 . . . . 102, 103

Bergman, Mark, San Francisco, CA (415) 391 9070 124

Bessett, Etienne, Ottawa, Canada (613) 748 3336 . . . . 216, 217

Bethke, Jennifer, San Francisco, CA (415) 551 1900 125

Bianchi, Natalie, Evanston, IL (847) 869 9820 168

Binney, Carrie, San Francisco, CA (415) 391 9070 124

Blackburn, John, London, England 171 734 7646 . . . .

. . . . 32, 33, 43, 56, 57, 69, 75, 81

Britton, Patti, Sonoma, CA (707) 938 8378 . . . . 64, 65, 66

Caldewey, Jeffrey, Napa, CA (707) 252 6666 . . . . 35, 36, 37, 74

Calkins, Mike, Seattle, WA (206) 467 5800 . . . . 178, 179

Callister, Mary Jane, New York, NY (212) 989 9153 . . . .

. . . . 71, 121, 132, 148, 149, 150, 151, 152, 184

Carson, Bill, Austin, TX (512) 371 1150 39

Casteix, Lane, New Orleans, LA (504) 484 7695 . . . . 53, 54

Celiz, Bob, San Francisco, CA (415) 495 8691 . . . . 204, 205

Ceradini, David, New York, NY (212) 255 7277 . . . . 55, 118, 136

Cerveny, Lisa, Seattle, WA (206) 467 5800 . . . . 178, 179

Chan, Alan, Hong Kong, China 852 2527 8228 134

Chang, Michael, Evanston, IL (847) 869 9820 247

Chavez, Marcos, New York, NY (212) 627 3200 . . . . 106, 107

Church, Stan, New York, NY (212) 755 2903 219

Ciesa, Lauren, East Lansing, MI (517) 351 2453 123

Clarke, Sally, New York, NY (212) 686 6046 27

Cochran, Bobbye, Chicago, IL (773) 404 0375 227

Corell, Karen, New York, NY (212) 375 0550 129

Corley-Macacy, Catherine, New Orleans, LA (504) 484 7695 53

Coverdale, Jac, Minneapolis, MN (612) 359 4387 22

Csaba, Almàssy, Budapest, Hungary 0036 1339 5348 169

Cuttione, Joe, New York, NY (212) 375 0550 129

Dalton, Katha, Seattle, WA (206) 467 5800 187

Davis, Grisson, Winston Salem, NC (336) 760 0770 . . . . 242, 243

Davison, Janice, San Francisco, CA (415) 495 8691 161

de Sibour, Peter, Minneapolis, MN (888) 590 4405 180

Decker, Steve, New Orleans, LA (504) 484 7695 54

Decker, Theo, Essen, Germany 49 201 842060 . . . . 101, 248

DeMarino, Anthony, New York, NY (212) 229 9377 126

Denat, Jean-Luc, Ottawa, Canada (613) 748 3336 . . . . 216, 217

Deutsch, Barry, San Francisco, CA (415) 487 8520 . . . .

. . . . 20, 21, 23, 72, 146, 182, 237

Devlin-Driskil, Patty, Santa Barbara, CA (805) 963 0734 . . . .

. . . . 24, 119, 130, 140164, 165

Dimeo, Joe New York, NY (212) 686 6046 27

Di Nardo, Franca, Toronto, Canada (416) 25 97834 137

Dorcas, John, Richardson, TX (972) 699 2143 112

Duckworth, Bruce, San Francisco, CA (415) 495 8691 . . . .

. . . . 16, 19, 161, 199, 204, 205

Duffy, Joe, Minneapolis, MN (612) 321 2333 185

Duggan, Belinda, London, England 44 171 734 7646 . . . . 32, 33, 75

Dunn, Chris, New York, NY (212) 683 7000 232

Eberhardt, Jared, Burlington, VT (802) 864 5884 235

Enoch, Tony, London, England 177 706 8762 44

Eplawy, Jason, Chicago, IL (312) 42 76413 233

Ernst, Tirza, Vernon Hills, IL (847) 573 8400 156

Evans, Leslie, Portland, ME (207) 874 0102 127

Farmer, Kaye, Seattle, WA (206) 467 5800 187

Fili, Louise, New York, NY (212) 989 9153 . . . .

. . . . 71, 121, 132, 148, 148, 150, 151, 152, 184

Fletcher, Victoria, London, England 171 470 4000 . . . . 170, 171

Florsheim, Alan, Seattle, WA (206) 467 5800 . . . . 178, 179, 187

Fornebo, Morten, Oslo, Norway 47 22 549500 52

Foshaug, Jackie, San Francisco, CA (415) 896 0499 . . . . 76, 77

Franklin, Dan, Portland, OR (503) 228 1381 155

Fuhrman, Marc, Evanston, IL (847) 869 9820 246

Glaister, Nin, New York, NY (212) 755 2903 . . . . 214, 219

Gorder, Genevieve, New York, NY (212) 981 7600 . . . . 2, 50, 51

Granzow, Gwen, Racine, WI (262) 639 2080 162

Greteman, Sonia, Wichita, KS (316) 263 1004 . . . .

. . . . 120, 158, 192, 195, 224, 225, 236

Gutwirth, Nat, Philadelphia, PA 141

Haggerty, Lawrence, New York, NY (212) 755 2903 219

Haislip, Heather, New York, NY (212) 805 2385 . . . . 87, 88, 221

Hale, Tim, Richardson, TX (972) 699 2143 . . . . 112, 113, 114, 115

Ham, Kathryn, New York, NY (807) 0248 175

Hames, Kyle, Minneapolis, MN (612) 339 5181 . . . . 196, 197

Harcus, Annette, Surry Hills, Australia 61 2 9212 2755 . . . .

. . . . 59, 60, 92, 98, 99, 210

Harris, Cabell, Richmond, VA (804) 358 9366 17

Hensley, Kelly, San Francisco, CA (415) 282 8678 145

Hinrichs, Kit, San Francisco, CA (415) 896 0499 . . . .

. . . . 76, 77, 142, 143, 206, 207

Hirano, Keiko, Toyko, Japan 81335725111 . . . . 94, 95, 108, 109

Hishida, Koichi, Osaka, Japan 816 6 6941 9618 73

Holly, Greg, Portland, OR (503) 872 8940 . . . . 67, 154

Hom, Deborah, San Diego, CA (619) 234 6631 . . . . 82, 83, 84, 85

Hom, Joanne, Berkeley, CA (510) 704 7500 . . . . 144, 167

Hook, Bryan, Edinburgh, Scotland 131 662 0662 166

Hornall, John, Seattle, WA (206) 467 5800 187

Horrigan, Michael, Ada, MI (616) 787 5168 . . . . 122, 174

Hough, Tom, Dallas, TX (214) 969 1050 . . . . 25, 147

Hubbard, Tom, Portland, ME (207) 874 0102 127

Ikeda, Shyuich Toyko, Japan 81335725111 . . . . 94, 95, 108, 109

Ikegaya, Junko Toyko, Japan 81335725111 . . . . 94, 95, 108, 109

Irvine, Steve, London, England 441 777 068 762 44

Jackman, Paul, Ada, MI (616) 787 5168 108 . . . . 122, 174

Jacobs, Brian, San Francisco, CA (415) 896 0499 . . . . 142, 143

Jacobs, Diane, Vernon Hills, IL (847) 573 8400 156

Jacobs, Nathalie, New York, NY (212) 979 8900 173

Jager, Michael, Burlington, VT (802) 864 5884 235

Janney, Dawn, San Francisco, CA (415) 487 8520 . . . . 20, 72, 146

Johnson, Haley, Richmond, VA (804) 358 9366 17

Jones, Kirk, San Francisco, CA (415) 882 9406 80
Judit, Toth, Budapest, Hungary 0036 1339 5348 160

Kaginada, Paul, San Francisco, CA (415) 255 0125 . . . . 63, 116, 117
Kaiser, Orville, New York, NY (212) 683 7000 232
Karlsson, Hjalti, New York, NY (212) 647 1789 191
Kikuch, Taisuke Toyko, Japan 81335725111 . . . . 94, 95, 108, 109
Kilmer, Richard, Albuquerque, NM (505) 232 6103 . . . . 135, 198
Kim, Aline, La Mirada, CA (562) 941 2090 215
Klim, Matt, Avon, CT (860) 678 1222 . . . . 47, 49
Klotnia, John, New York, NY (212) 683 7000 232
Kohla, Hans, Surrey Hills, Australia 61 2 9212 2755 59
Kohlman, Gary, Albuquerque, NM (505) 232 6103 135
Kudo, Aoshi Toyko, Japan 81335725111 . . . . 94, 95, 96, 97, 108, 109
Kulisek, Christine, New York, NY (212) 375 0550 129

L'Écuyer, Mario, Ottawa, Canada (613) 748 3336 216
Lagenecker, Jackie, Racine, WI (262) 639 2080 162
Ledgerwood, Lisa, Phoenix, AZ (602) 954 6992 . . . . 93, 172
Ledgerwood, Mark, Phoenix, AZ (602) 954 6992 . . . . 93, 172
LeJeune, Sharon, Dallas, TX (214) 826 4631 110
Leong, David, Berkeley, CA (510) 704 7500 167
Lillerik, Kristine, Oslo, Norway 47 22 549500 52
Lindgren, Mitchell, Minneapolis, MN (612) 339 1595 . . . . 18, 131, 208
Liska, Steve, New York, NY (212) 627 3200 . . . . 106, 107
Lloyd, Mark, London, England 177 706 8762 44
Lo, Peter, Hong Kong, China (852) 2527 8228 134
Loder, Tom, Philadelphia, PA 141
Lucas, John, San Francisco, CA (415) 487 8520 182
Luk, Anthony, Berkeley, CA (510) 704 7500 144

Mannes, Todd, Minneapolis, MN (888) 590 4405 180
Markus, Craig (212) 981 7600 . . . . 2, 50, 51
Marshall, Randall, Albuquerque, NM (505) 232 6103 . . . . 135, 198
Martin, Fletch, Chicago, IL (312) 427 6413 233
Mastandrea, Mary Anne (415) 538 8130 26
Matsumoto, Izumi, Toyko, Japan 81335725111 . . . . 94, 95, 108, 109
Matsumoto, Takaaki, New York, NY (212) 807 0248 175
Matsunaga, Shin, Tokyo, Japan 81 3 5353 0170 . . . . 29, 30, 31, 222
Max, Sonja, Seattle, WA (206) 467 5800 . . . . 130, 157
Maxwell, Red, Winston Salem, NC (336) 760 0770 . . . . 242, 243
McGowan, Rich, Minneapolis, MN (612) 339 1595 . . . . 18, 131, 208
Mecchi, Michelle, Saint Helena, CA (707) 963 4152 . . . . 62, 133
Michels, Nicole, New York, NY (212) 229 9377 126
Miguel, Allison, London, England 171 470 4000 . . . . 170, 171
Milostic, Mario, Sydney, Australia 02 9929 2458 . . . . 45, 92, 98, 211
Mortensen, Steve, Palo Alto, CA (650) 321 8138 . . . . 240, 241
Murawski, Bill, Torrance, CA (310) 329 0800 . . . . 46, 48

Needham, Nick, Edinburgh, Scotland 1316620662 166
Nestor, Okey, New York, NY (212) 229 9377 126
Nicosia, Joan, New York, NY (212) 375 0550 129
Nishi, Jana, Seattle, WA (206) 467 5800 . . . . 130, 157
Novello, Peter, New York, NY (212) 255 7277 136

Oates, Roberta, London, England 171 734 7646 . . . . 56, 57, 69, 81
Okumura, Akio, Oska, Japan 816 6 6941 9618 . . . . 42, 73
Oliver, Mark, Santa Barbara, CA (805) 963 0734 . . . .
. . . . 24, 38, 119, 130, 140, 164, 165
Olofsson, Håkan, Stockholm, Sweden 46 858 70 8000 78
Olofsson, Jörgen, Stockholm, Sweden 46 858 70 8000 78
Osborne, Michael, San Francisco, CA (415) 255 0125 . . . .
. . . . 63, 116, 117, 188, 189

Pagoda, Carlo, San Francisco, CA (415) 551 1900 125
Pan, Bernice, New York, NY (212) 334 6151 223
Park, Amanda, Torrance, CA (310) 329 0800 . . . . 46, 48
Perez, Miguel, San Diego, CA (619) 234 6631 . . . . 181, 193, 194, 213
Peterson, Anne, Minneapolis, MN (612) 338 4462 105

Piatek, Keri, New York, NY (212) 255 7277 55
Pierce, Brenna, Santa Barbara, CA (805) 963 0734 . . . . 38, 140
Pino, Eric, San Francisco, CA (415) 487 8520 182
Piper-Hauswirth, Todd, Minneapolis, MN (612) 339 5181 . . . . 196, 197
Pirtle, Woody, New York, NY (212) 683 7000 232

Radtke, Steve, Milwaukee, WI (414) 228 9666 209
Ravlet, Allen, San Francisco, CA (415) 495 8691 16
Raymer, Lori, New York, NY (212) 255 7277 55
Reeves, Patrick, Richardson, TX (972) 699 2143 115
Regenbogen, Michelle, San Francisco, CA (415) 255 0125 . . . . 188, 189
Riddle, Tom, Minneapolis, MN (612) 321 2333 185
Roberts, Sarah, London, England 171 734 7646 43
Rossouw, Jacques, San Francisco, CA (415) 487 8520 . . . . 23, 182, 237
Ryan, Melonie, Surrey Hills, Australia 61 2 9212 2755 . . . . 60, 99

Sagmeister, Stefan, New York, NY (212) 647 1789 191
Samaripa, Jeff, San Diego, CA (619) 234 6631 183
Sánchez, Lucero, Ottawa, Canada (613) 748 3336 217
Sapp, Ingrid, Salzburg, Austria 43 662 840 440 . . . . 111, 138, 139
Sayles, John, Des Moines, IA (515) 279 2922 100
Schreiber, Curt, Chicago, IL (312) 427 6413 233
Scott, Fiona, Edinburgh, Scotland 131 662 0662 166
Scott, Julie, Phoenix, AZ (602) 954 6992 172
Sempi, Enrico, Novara, Italy 39 0321 35662 . . . . 230, 231
Serrano, Jose, San Diego, CA (619) 234 6631 . . . .
. . . . 82, 83, 84, 85, 181, 183, 193, 194, 213, 234
Sibley, Don, Dallas, TX (214) 969 1050 . . . . 25, 147
Sidie, Ingred, Kansas City, MO (816) 472 8668 200
Sikora, Steven, Minneapolis, MN (612) 338 4462 . . . . 105, 159
Smiler, Ellen, Los Angeles, CA (310) 208 5645 203
Smith, Brian, Toronto, Canada (416) 259 7834 41
Smith, Donna, Ada, MI (616) 787 5168 122
Sonderegger, Michelle, Kansas City, MO (816) 472 8668 200
Sousa, Gabriella, Yoronto, Canada (416) 259 7834 41
Stair, Jessica, Winston Salem, NC (336) 760 0770 . . . . 242, 243
Stanard, Michael, Evanston, IL (847) 869 9820 . . . . 168, 245, 246, 247
Steimel, Keith, New York, NY (212) 686 6046 . . . . 27, 128, 238, 239
Sterling, Jennifer, San Francisco, CA (415) 612 3481 . . . . 90, 91
Stermole, Rick, New York, NY (212) 229 9377 126
Stigler, Bruce, Seattle, WA (206) 467 5800 187
Stimel, Keith, New York, NY (212) 686 6046 . . . . 128, 226
Strange, James, Wichita, KS (316) 263 1004 . . . .
. . . . 120, 158, 192, 195, 224, 225, 236
Stumpf, Ralf, Essen, Germany 49 201 842060 186
Sudduth, Toby, San Francisco, CA (415) 551 1900 125
Sudman, Sharon, Minneapolis, MN (612) 339 1595 208
Sundermann, Michael, East Lansing, MI (517) 351 2453 123
Svensson, Mats, Stockholm, Sweden 46 858 70 8000 78

Tanner, Ellen, Richardson, TX (972) 699 2143 . . . . 114, 115
Tanaka, Yasuo, Osaka, Japan 816 6 6675 0118 . . . . 228, 229
Tebon, Marc, Milwaukee, WI (414) 228 9666 209
Thoelke, Eric, Saint Louis, MO (314) 588 1500 . . . . 201, 202
Ting, Phillip, San Francisco, CA (415) 391 9070 124
Trevisan, Antonella, Novara, Italy 39 0321 35662 . . . . 230, 231
Turner, David, San Francisco, CA (415) 495 8691 . . . .
. . . . 16, 19, 161, 199, 204, 205

Ueno, Mitsuo, Oska, Japan 816 6 6941 9618 42
Ulrich, Lowthar, Yoronto, Canada (416) 25 97834 137
Upton, Bob, Minneapolis, MN (612) 359 4387 22

Van Noy, Jim, Yorrance, CA (310) 329 0800 . . . . 46, 48
Van Wyck, Chris, East Lansing, MI (517) 351 2453 123
Vandekerckhove, Kristy, Evanston, IL (847) 869 9820 245
Voll, Linda, Chicago, IL (312) 943 5995 163
Voss, Axel, Essen, Germany 49 201 842060 248

Watts, Peter 212

Waters, Mark, San Francisco, CA (415) 495 8691 161

Wharton, Paul, Minneapolis, MN (888) 590 4405 180

White, Shoshannah, Portland, ME (207) 874 0102 127

Wilson, Lori B., Dallas, TX (214) 826 4631 110

Wynn, Lori, San Francisco, CA (415) 487 8520 . . . . 20, 21, 23, 182, 237

Yamamoto, Naomi, New York, NY (212) 805 2385 . . . . 87, 88, 221

Zastoupil, Chris, Minneapolis, MN (888) 590 4405 180

Zhang, Steven, Richardson, TX (972) 699 2143 . . . . 113, 114, 115

Zoltan, Halasi, Budapest, Hungary 361 339 5348 153

PhotographersIllustratorsCopywriters

Aman, Russ, Richardson, TX (972) 699 2143 . . . . 112, 113, 114, 115

Anderson, Charles S., Minneapolis, MN (612) 339 5181 . . . . 196, 197

Anderson, Tim, Winston Salem, NC (336) 760 0770 . . . . 242, 243

Arnold, Keith, Surry Hills, Australia, 61 2 9212 2755 . . . . 60, 99

Baxter, Scott, Phoenix, AZ (602) 954 6992 93

Beecham, Greg , San Francisco, CA (415) 487 8520 23

Bessette, Etienne, Ottawa, Canada (613) 748 3336 215

Bianchi, Natalie, Evanston, IL (847) 869 9820 168

Bishop, David, Saint Helena, CA (707) 963 4152 . . . . 63, 133

Block, Jamie, New York, NY (212) 647 1789 191

Britton, Patti, Sonoma, CA (707) 938 8378 . . . . 64, 65, 66

Brown, Calef, Dallas, TX (214) 969 1050 23

Burke/Triolo, Santa Barbara, CA (805) 963 0734 . . . . 164, 165

Butler, Erik, San Francisco, CA (415) 538 8130 26

Calero, Adolfo, San Francisco, CA (415) 282 8678 145

Carson, Bill, Austin, TX (512) 371 1150 39

Chang, Michael, Evanston, IL (847) 869 9820 247

Cochran, Bobbye, Chicago, IL (773) 404 0375 227

Craig, Dan, Santa Barbara, CA (805) 963 0734 . . . . 164, 165

Csaba, Almàssy, Budapest, Hungary, 0036 1339 5348 169

Deahl, David, San Diego, CA (619) 234 6631 183

Delavison, Justin, San Francisco, CA (415) 495 8691 199

DNA Illustrations, New Orleans, LA (504) 484 7695 54

Donavan, Michael, Surry Hills, Australia, 61 2 9212 2755 . . . . 59, 92

Dorcas, John, Richardson, TX (972) 699 2143 112

Draper, David, London, England, 44 171 734 7646 . . . . 32, 33

Driskel, Patty, Santa Barbara, CA (805) 963 0734 24

Duke, Larry, San Francisco, CA (415) 487 8520 21

Ehrbar, Barbara, New York, NY (212) 647 1789 191

Erdman, Michael, San Francisco, CA (415) 551 1900 125

Erickson, Jim, Minneapolis, MN (512) 338 4462 159

Esner, Franz , Salzburg, Austria, 43 662 840440 . . . . 138, 139

Fenton, Simon, Surry Hills, Australia, 61 2 92122755 98

Foard, Lane, San Francisco, CA (415) 538 8130 26

Fonseca, Caio, San Francisco, CA (415) 255 0125 63

Franklin, Dan, Portland, OR (503) 228 1381 155

Grashen, James, New York, NY (212) 989 9153 151

Greteman, Sonia , Wichita, KS (314) 236 1004 236

Hamagami Carroll & Associates, Berkeley, CA (510) 704 7500 167

Hames, Kyle, Minneapolis, MN (612) 339 5181 . . . . 196, 197

Hansen, Clint. New York, NY (212) 686 6046 27

Hards, Stephen J., Ottawa, Canada, 613 748 3336 . . . . 216, 217

Harms, Deanna, Wichita, KS (316) 263 1004 . . . . 235, 236

Headlight Innovative Imagery, Ottawa, Canada, 613 748 3336 217

Holownia, Tom, Berkeley, CA (510) 704 7500 144

Hough, Tom, Dallas, TX (214) 969 1050 147

Howgil, Jody, San Francisco, CA (415) 487 8520 72

Hutchison, Bruce, Portland, ME (207) 874 0102 127

Ingalls, Randall, San Francisco, CA (415) 391 9070 124

Iridium Marketing & Design, Ottawa, Canada 613 748 3336 215

Jarvis, John, Minneapolis, MN (612) 321 2333 185

Kelley, Gary, New York, NY (212) 989 9153 121

Kim, Julie, New York, NY (212) 334 6151 223

Kimball, Anton, Portland, OR (503) 872 8940 154

Kirk, Jon, Richardson, TX (972) 699 2143 114

Kirkpatrick, Amy, Minneapolis, MN (512) 338 4462 159

Klim, Greg, Avon, CT (860) 678 1222 . . . . 47, 49

Krause, Peter, Minneapolis, MN (612) 359 4387 22

Kuraoka, John, San Diego, CA (619) 234 6631 234

Laire, Richard, San Francisco, CA (415) 612 3481 . . . . 90, 91

Lam, Ed, San Francisco, CA (415) 487 8520 146

Landry, Cat, Torrance, CA (310) 329 0800 46

Lawrence, John, Santa Barbara, CA (805) 963 0734 140

Leatherman Tool Group 178, 179

Leman, Martin, London, England, 171 734 7646 69

Lewis, Gretchen, Berkeley, CA (510) 704 7500 144

Lindgren, Cindy, Minneapolis, MN (612) 339 1595 131

Luk, Anthony, Berkeley, CA (510) 704 7500 144

MacDonald, Kathryn, San Francisco, CA (415) 282 8678 145

Mari, Edoardo, Novara, Italy, 39 0321 35662 230

Marsh, James, London, England, 44 171 734 7646 75

Marshall, Randall, Albuquerque, NM (505) 232 6103 135

Mastandrea, Mary Anne (415) 538 8130 26

Matsumoto, Takaaki, New York, NY (212) 807 0248 175

May, Andrea, San Diego, CA (619) 234 6631 . . . . 83, 84, 85

Merrell, Patrick, Vernon Hills, IL (847) 573 8400 156

Michels, Nicole, New York, NY (212) 229 9377 126

Miller, Cynthia, Minneapolis, MN (512) 338 4462 159

Milostic, Mario, Sydney, Australia, 02 9929 2458 45

Murphy, Tara, Evanston, IL (847) 869 9820 168

Newbold, Greg, Santa Barbara, CA (805) 963 0734 38

Newton, Paul, Surry Hills, Australia, 61 2 92122755 99

Norwell, Jeff, San Francisco, CA (415) 487 8520 23

Odenhansen, DetlefEssen, Germany, 49 201 842060 248

Oliver, Mark, Santa Barbara, CA (805) 963 0734 . . . .
. . . . 24, 38, 119, 130, 140, 164, 165

O'Neal, David, New York, NY (212) 686 6046 27

Oughton, Taylor , San Francisco, CA (415) 487 8520 23

Padavos, Lefteris, Torrance, CA (310) 329 0800 . . . . 46, 48

Patch, Gary, Minneapolis, MN (612) 338 4462 105

Patteson, Richard, London, England, 441 777 068 762 44

Pelaun, Daniel, San Francisco, CA (415) 896 0499 . . . . 76, 77

Pemrick, Lisa, Minneapolis, MN (612) 339 5181 . . . . 196, 197

Perez, Miguel, San Diego, CA (619) 234 6631 183

Petrauskas, Kathy, New York, NY (212) 255 7277 55

Pierce, Brenna, Santa Barbara, CA (805) 963 0734 119

Potter, Nicole, Surry Hills, Australia, 61 2 9212 2755 59

Powell, Pam, Minneapolis, MN (888) 590 4405 180

Quaranta, Sergio, Novara, Italy 39 0321 35662 230

Reeves, Patrick, Richardson, TX (972) 699 2143 115
Resource Games, Seattle, WA (206) 467 5800 157
Riddle, Tom, Minneapolis, MN (612) 321 2333 185
Rose, Peter, Portland, OR (503) 228 1381 155
Rownd, Jim, Minneapolis, MN (612) 339 1595 18
Ryan, Melonie, Surry Hills, Australia, 61 2 9212 2755 60

Sabin, Tracy, New York, NY (212) 755 2903 . . . . 181, 183, 184, 213
Sainz, Roger, New York, NY (212) 255 7277 118
Salerno, Steve, Santa Barbara, CA (805) 963 0734 130
Sapp, Ingrid, Salzburg, Austria (43 662 840 440 . . . . 111, 138, 139
Sater, Arild, Oslo, Norway, 47 22 549500 52
Sayles, John, Des Moines, IA (515) 279 2922 100
Schuchman, Bob, Torrance, CA (310) 329 0800 48
Sonderegger, Michelle, Kansas City, MO (816) 472 8668 200
Stahl, Nancy, San Diego, CA (619) 234 6631 181
Starr, Jim, Philadelphia, PA 141
Stava, Susan, New York, NY (212) 647 1789 191
Steinbrenner, Karl, Richmond, VA (804) 358 9366 17
Sticky Fingers Bakery, Seattle, WA (206) 467 5800 130

Stradling, Bob, London, England, 441 777 068 762 44
Strange, James, Wichita, KS (316) 263 1004 . . . . 224, 225, 236

Takagi, Mike, Vernon Hills, IL (847) 573 8400 156
Tanner, Ellen, Richardson, TX (972) 699 2143 . . . . 114, 115
Trewartha, Kelly , Minneapolis, MN (612) 359 4387 22

Ulrich, Lowthar, Toronto, Canada (416) 259 7834 41

Vanderschuit, Carl, San Diego, CA (619) 234 6631 . . . . . . 82, 83, 84, 85, 234
Veassey, Nick, London, England 171 470 4000 . . . . 170, 171
Vichitlakakarn, Ekasit, New York, NY (212) 255 7277 136

Wakely, David, San Francisco, CA (415) 255 0125 . . . . 188, 189
Walker, Jeff, Chicago, IL (312) 42 76413 233
Ward, Brion, Sonoma, CA (707) 938 8378 . . . . 64, 65
Weir, Denise, Seattle, WA (206) 467 5800 157
Wepplo, Mike, New York, NY (212) 375 0550 129
Wetendorf, Catherine, Philadelphia, PA 141
Wilson, Michael, Dallas, TX (214) 826 4631 110
Wu, Leslie, Chicago, IL (312) 943 5995 163

Zhang, Steven, Richardson, TX (972) 699 2143 . . . . 113, 115
Zorn, Boris, Essen, Germany 49 201 842060 186